Green Finance and Investment

Overcoming Barriers to International Investment in Clean Energy

OECD
BETTER POLICIES FOR BETTER LIVES

This work is published under the responsibility of the Secretary-General of the OECD. The opinions expressed and arguments employed herein do not necessarily reflect the official views of OECD member countries.

This document and any map included herein are without prejudice to the status of or sovereignty over any territory, to the delimitation of international frontiers and boundaries and to the name of any territory, city or area.

Please cite this publication as:
OECD (2015), *Overcoming Barriers to International Investment in Clean Energy*, Green Finance and Investment, OECD Publishing, Paris.
http://dx.doi.org/10.1787/9789264227064-en

ISBN 978-92-64-22704-0 (print)
ISBN 978-92-64-22706-4 (PDF)

Serie: Green Finance and Investment
ISSN 2409-0336 (print)
ISSN 2409-0344 (online)

The statistical data for Israel are supplied by and under the responsibility of the relevant Israeli authorities. The use of such data by the OECD is without prejudice to the status of the Golan Heights, East Jerusalem and Israeli settlements in the West Bank under the terms of international law.

Photo credits: Cover © mika48/Shutterstock.com.

Corrigenda to OECD publications may be found on line at: *www.oecd.org/about/publishing/corrigenda.htm*.

Foreword

Limiting the extent of climate change requires substantial and sustained reductions in greenhouse gas emissions. Emissions from fossil-fuel use currently account for around two-thirds of total emissions. Investment in clean energy therefore needs to be scaled up significantly if we are to limit the severe risks we face from climate change. Mobilising clean-energy investment also creates benefits and opportunities for developed and developing countries alike, such as: reduced local air pollution and associated health costs; cost-effective access to energy in rural and remote areas; improved energy security and reduced reliance on fossil-fuel imports; and technology transfer and innovation across the value chain.

Public action will have to focus on ways of increasing private investment in clean-energy systems at an unprecedented pace and on a much larger scale than under a "business-as-usual" scenario. Governments have a critical role to play in providing the right policy environment that can enhance the risk-return profile of clean-energy investments. This topic was examined in a previous OECD report, the *Policy Guidance for Investment in Clean Energy Infrastructure*, which was annexed to the Communiqué of G20 Finance Ministers and Central Bank Governors at their meeting of 10-11 October 2013.

Governments can also provide targeted support to leverage private investment in clean energy. In fact, largely driven by public incentives, clean-energy investment – especially in solar photovoltaic (PV) and wind energy – has increased significantly over the past decade. In addition, several countries have removed trade and investment barriers to clean energy, such as import tariffs and restrictions on foreign direct investment. As a result, international trade and investment have played a major role in driving significant cost reductions in the solar-PV and wind-energy sectors.

Besides these favourable developments, another trend involving possibly less beneficial impacts has emerged since the 2008 financial crisis. The perceived potential of clean energy to act as a lever for growth and employment has led several OECD and emerging economies to design green industrial policies aimed at encouraging or protecting domestic solar panel and wind-turbine manufacturers, notably through setting local-content requirements. The report *Overcoming Barriers to International Investment in Clean Energy* documents the increasing use of such local-content requirements and assesses their impact across the global solar- PV and wind-energy value chains. It provides new econometric evidence that local-content requirements have hindered investment in solar PV and wind energy. At the same time, recent country experiences suggest that the impact of such requirements on local job creation and value added is at best mixed. The report also takes stock of other measures that may hamper international investment in solar PV and wind energy.

Creating policy frameworks that enable international trade and investment flows in clean energy is crucial. Policies that promote open, competitive and demand-driven markets for clean energy will support the continued cost reductions needed for a cost-effective transition to a low-carbon energy system, reducing the amount of public incentives needed to scale up the deployment of clean technologies.

The OECD is thus proud to contribute this report on *Overcoming Barriers to International Investment in Clean Energy* as part of the evidence-base needed to improve the coherence and cost-effectiveness of clean-energy support policies, including trade and investment policies. By assessing the impacts of local-content requirements across different segments of the solar-PV and wind-energy value chains, this work provides policy makers with the necessary evidence to guide the design of clean-energy support policies.

Angel Gurría
OECD Secretary-General

Acknowledgements

This report was jointly developed by the OECD Investment Committee and the Working Party on Climate, Investment and Development (WPCID) of the Environment Policy Committee (EPOC). It is a joint endeavour of the Investment Division of the OECD Directorate for Financial and Enterprise Affairs, and the Climate, Biodiversity and Water Division of the OECD Environment Directorate. The report benefited from substantial contributions from the Working Party of the Trade Committee and the Joint Working Party on Trade and Environment (JWPTE) of EPOC and the Trade Committee. It builds on relevant OECD work and instruments developed by the Investment and Environment Policy Committees, as well as the Competition, Corporate Governance and Trade Committees, the JWPTE, and the International Energy Agency (IEA).

This report was co-authored by Geraldine Ang, Karim Dahou and Robert Youngman, with substantial input received from Ronald Steenblik from the Trade Directorate. Miguel Cárdenas Rodríguez from the Environment Directorate authored the econometric analysis and Chung-a Park from the Directorate for Financial and Enterprise Affairs prepared the 2014 OECD Investor Survey. The authors are thankful to their colleagues at the OECD Secretariat and IEA who provided valuable comments and expert review, including: Heymi Bahar, Manuel Baritaud, Richard Baron, Adrian Blundell-Wignall, Maria Borga, Nils-Axel Braathen, Simon Buckle, Antonio Capobianco, Juan Casado Asensio, Hans Christiansen, Anthony Cox, Jane Ellis, Justine Garrett, David Gaukrodger, Nathalie Girouard, Christopher Kaminker, Osamu Kawanishi, Emilie Kothe, Iza Lejarraga, James Messent, Sebastien Miroudot, Fernando Mistura, Ana Novik, Pierre Poret, Andrew Prag, Jehan Sauvage, Susan Stone, Sara Sultan, Monika Sztajerowska and Žiga Zarnic. The authors are grateful for valuable assistance provided by Lesley O'Sullivan, Jade Baker and Sama Al-Taher as well as Andrew Esson, Katherine Kraig-Ernandes and Janine Treves who assisted with editing and typesetting and prepared the manuscript for publication.

The authors would like to thank the following experts and peer reviewers for their valuable inputs, comments and guidance: Peter Brun (SETI Alliance), Giorgio Corbetta (EWEA), Jonas Kasteng (Swedish National Board of Trade), Angeliki Koulouri (EWEA), Jacopo Moccia (EWEA), Ivan Pineda (EWEA) and Kenneth G. Ruffing.

The report draws on an extensive review of the literature, new empirical evidence and stakeholder consultations. In particular, it incorporates feedback received during a roundtable consultation held on 6 December 2013 at the OECD to gather insights from key stakeholders on good practices to address policy impediments to international investment in solar and wind energy (OECD, 2013h). The authors would particularly like to thank the following participants for their input: Erich Becker (Zouk Capital LLP), Denis Cochet

(Alstom Power), Jose Galindez (Solarpack), Check Haris (Alstom Power), Janina Ketterer (EBRD), Karsten Loeffler (Allianz Climate Solutions), Pierre Tardieu (EWEA), Florian Wessendorf (VDMA German Engineering Federation), Simon Wilde (Macquarie Capital), and Andrea Wuttke (Canadian Solar).The authors would also like to thank Ethan Zindler (Bloomberg New Energy Finance, BNEF) and the participants of another stakeholder consultation held at the 2013 BNEF Summit in New York. Finally, the authors would like to extend special thanks to survey participants who took the time to complete the 2014 Investor Survey on "Achieving a Level Playing Field for International Investment in Clean Energy". The authors are also grateful to SETI Alliance, European Wind Energy Association (EWEA), European Photovoltaic Industry Association (EPIA) and Solar Energy Industries Association (SEIA) for circulating the 2014 OECD Investor Survey to their members.

Table of contents

Acronyms and abbreviations

AD	anti-dumping duty
AFASE	Alliance for Affordable Solar Energy
AGID	Advisory Group on Investment and Development
AHJ	Authority Having Jurisdiction (Canada)
APEC	Asia-Pacific Economic Cooperation
APP	Permanent Preservation Areas
AWEA	American Wind Energy Association
BIAC	Business and Industry Advisory Committee to the OECD
BNDES	Brazilian Development Bank
BNEF	Bloomberg New Energy Finance
BTA	bilateral trade agreement
CAWEA	Canadian Wind Energy Association
CDM	Clean Development Mechanism
CERC	Central Electricity Regulatory Commission
CGE	Computable General Equilibrium
CSA	Canadian Standards Association
c-Si	crystalline silicon
CTG	Council for Trade in Goods
CVD	countervailing duty
CWET	Centre for Wind Energy Technology (India)
CO_2	carbon dioxide
DCR	domestic content requirement
DNV	Det Norske Veritas
DSB	Dispute Settlement Body
ECA	export credit agencies
EIA	environmental impact assessment

EPIA	European Photovoltaic Industry Association
ERP	Energy for Renewable Power
ESA	environmental site assessment
EU	European Union
EWEA	European Wind Energy Association
Ex-Im Bank	US Export Import Bank
FDI	Foreign Direct Investment
FiT	feed-in tariff
GATT	General Agreement on Tariffs and Trade
GHG	greenhouse gas
GL	Germanischer Lloyd
GPA	Government Procurement Agreement
GVC	global value chain
ICTSD	International Centre for Trade and Sustainable Development
IEA	International Energy Agency
IEEE	Institute of Electrical and Electronics Engineers
IEC	International Electrotechnical Commission
IPP	independent power producers
IRENA	International Renewable Energy Agency
ISO	Industrial Organization for Standardization
JNNSM	Jawaharal Nehru National Solar Mission (India)
JWPTE	Joint Working Party on Trade and Environment (OECD)
LCR	local-content requirement
M&A	merger and acquisition
MFN	most favoured nation
MW	mega watt
NERC	North American Electricity Reliability Corporation
NERSA	National Energy Regulator of South Africa
NRDC	Natural Resources Defense Council
OEM	Original Equipment Manufacturers
PROINFA	Programme of Incentives for Alternative Electricity Sources (Brazil)
PV	photovoltaic

R&D	research and development
RAS	Simplified Environmental Report
REN21	Renewable Energy Policy Network for the 21st Century
REQ	renewable energy quota
RTA	regional trade agreement
SCM	subsidies and countervailing measures
SETI	Sustainable Energy Trade Initiative
SOE	state-owned enterprise
TBT	technical barriers to trade
TDI	trade defence instruments
TiVA	Trade in Value Added database (OECD)
TRIMs	Trade-Related Investment Measures
UNCTAD	United Nations Conference on Trade and Development
USW	United Steel Workers
WPCID	Working Party on Climate, Investment, and Development (OECD)
WTO	World Trade Organization
WTTC	Wind Tower Trade Coalition

Executive summary

Investment in clean energy needs to be mobilised at pace and at scale to contribute to mitigating climate change and achieve the transition to a low-carbon energy system. Doing so can also provide numerous other social and economic benefits for both developed and developing countries. "Clean energy" as defined in this paper includes the following sectors: solar, wind, small and large hydroelectric, geothermal, marine, biomass and waste-to-energy power plants, carbon capture and storage (CCS) technologies and energy-efficient technologies such as smart grids and electric vehicles. Scaling up investment in clean energy will require the mobilisation of private investment from both domestic and international sources and policies that provide a supportive domestic investment environment.

Over the past decade, governments have provided substantial incentives to the deployment of clean energy, in addition to removing trade and investment barriers, such as import tariffs and restrictions to foreign direct investment (FDI). Investment in the solar photovoltaic (PV) and wind-energy sectors grew dramatically between 2004 and 2011, before declining in 2012-13 as a result of excess capacity, market consolidation and declining costs.

International trade and investment have also played an important role in driving the growth of the solar-PV and wind-energy sectors and their integration into global value chains. In particular, two trends have important implications for policy makers:

- Production and activities in the solar-PV and wind-energy sectors are increasingly reliant on imported intermediate inputs. Policies aimed at protecting domestic manufacturers may thus hinder the profitability of downstream activities, e.g. by raising the cost of inputs.
- The share of downstream activities in value added, employment and investment has increased relative to midstream activities such as manufacturing. Thus, policies targeting upstream and midstream manufacturing activities may not be effective in creating domestic jobs and value across the value chains.

This report focuses on the rise and impact of local-content requirements (LCRs) in the solar-PV and wind-power industries. In the post-crisis recovery context, the perceived potential of clean energy to support growth and employment has led several governments to implement green industrial policies aimed at protecting domestic solar-panel and wind-turbine manufacturers. In particular, governments have established LCRs to facilitate public acceptance of policy support to clean energy, as they are commonly thought to create value-added and jobs locally. Local-content requirements (or "locally manufactured content requirements") for solar or wind energy typically require developers and investors, in order to be eligible for policy support or tenders, to source a specific share of manufactured components or equipment locally. Local-content requirements have been planned or implemented in solar and wind energy in at least 21 countries, including 16 OECD countries and emerging economies, mostly since 2009. This has prompted five World Trade Organization (WTO) disputes since 2010.

In a context of global value chains, new empirical evidence provided in this report shows that LCRs can hamper international investment in solar- and wind-energy generation in the country that adopts them and globally. In addition, experience from recent country case studies suggests that LCRs have mixed or negative impacts on local job creation, value added and technology transfer in solar PV and wind energy when the full value chain is taken into account. By raising the cost of inputs for downstream businesses, LCRs can lead to increased overall costs, reduced price competitiveness, less international investment, and higher wholesale electricity prices.

Although this report recognises the reasons why governments have adopted local-content requirements, it advises against their implementation. To avoid negative impacts, policy makers should design domestic incentive measures that do not differentiate between domestic and international investors. Possible options include: well-targeted R&D support, which can stimulate innovation across segments of the value chains, build local manufacturing capability and encourage technology transfer from imports and FDI; training programmes and promotion measures, to improve the technological skills of manufacturers, build local capability of downstream firms and encourage innovation; and demand-side instruments, or more cost-effective carbon pricing instruments, to increase domestic demand and eventually support domestic manufacturing.

Foreign Direct Investment regulatory restrictions in solar PV and wind energy remain relatively limited in OECD countries. However, technical restrictions exist, such as divergent technical standards in selected wind-energy markets. Such policies can hinder international investment in wind energy by raising transaction costs, and reducing economies of scale.

While applied tariffs remain relatively low in solar and wind energy, the alleged use of actionable subsidies and dumping has led to a proliferation in retaliatory trade remedies – mostly by developed countries, but increasingly from emerging economies. Since 2009, governments have imposed nine anti-dumping duties (ADs) and seven countervailing duties (CVDs) on products associated with solar PV and wind energy, and launched 15 WTO investigations for ADs and nine investigations for CVDs (updated as of September 2014).

Creating a stable and predictable policy environment for both domestic and international investment in clean-energy generation is critical, as emphasised by the OECD *Policy Guidance for Investment in Clean Energy Infrastructure*. Supporting open, competitive and demand-driven solar-PV and wind-energy sectors would help sustain the trend towards cost reductions and make renewable energy more competitive vis-à-vis fossil-fuel energy. This, in turn, would reduce the cost of policy support to clean energy. Evidence-based analysis is needed to improve the coherence of clean-energy support policies and reduce their cost. International co-operation is also needed to align trade and investment policy in clean energy, including to responding to the escalation of local-content requirements.

Chapter 1

Achieving a level playing field for international investment in solar and wind energy

This chapter provides an integrated overview of the structure and scope of the report, which: 1) takes stock of outstanding barriers to international investment in solar PV and wind energy in OECD countries and emerging economies, with a focus on local-content requirements; and 2) assesses the impacts of local-content requirements on international investment in solar PV and wind energy, in a context of global value chains. This chapter also provides an overview of the rationale for scaling up investment in clean energy and summarises key implications of the report's findings for policy makers.

The need to scale up domestic and international investment in clean energy[1]

Investment in clean energy needs to be scaled up significantly in the coming years to contribute to mitigating climate change, and to ensure that the goal of holding global warming below two degrees (2°C)[2] is achieved. Addressing climate change and meeting the 2°C goal will require moving toward the complete elimination of net greenhouse gas (GHG) emissions from fossil-fuel combustion to the atmosphere in the second half of the century. However, the challenge of meeting rising energy demand continues to be mostly met by fossil fuels. Globally, fossil fuels still account for 79% of final energy consumption, a share that is almost unchanged since 1990. To address climate change, investments must be shifted from carbon-intensive to low-carbon, clean-energy infrastructure (OECD, 2015). A broad range of policy interventions are needed to transform the energy system, shift investment away from fossil fuels towards clean energy, and produce the requisite and unprecedented level of economic, social and technological change to achieve a 2°C goal (OECD, 2013i). The IEA estimates that cumulative investment of USD 53 trillion in energy supply and energy efficiency will be needed by 2035 to get the world on a path to achieve a two-degree goal, compared with USD 48 trillion based on today's policies (IEA, 2014c). Given the current strains on public finances, mobilising investment in clean energy will necessitate the leveraging of both domestic and international private investment.

In addition to reducing GHG emissions, actions to support investment in clean-energy infrastructure provide other benefits and economic opportunities for developed and developing countries. Opportunities include: "leapfrogging" existing fossil-fuel-based technologies to avoid locking-in carbon-intensive development pathways; facilitating cost-effective access to decentralised energy in rural and remote areas; reducing local air pollution and associated health costs; and stimulating innovation and technology transfer through international trade and investment. One specific action to support clean-energy investment – fossil-fuel subsidy reform – can improve the balance of payments and energy security for countries who import fossil fuels by reducing reliance on these imports. It also can relieve pressure on national budgets in countries that subsidise fossil fuels.[3] Investments in clean-energy infrastructure can also contribute to creating value added and employment at different stages of clean-energy sector value chains, including in downstream activities (such as project development, construction, installation, operations and maintenance) and midstream manufacturing activities. Through enhancing synergies between economic growth and climate change mitigation, investment in clean-energy infrastructure can help reduce the trade-offs between development and environmental priorities in the short run, and reduce costs in the long run.

Largely driven by government incentives, new investment[4] in clean-energy generation increased six-fold between 2004 and 2011, reaching USD 279 billion in 2011 (FS-UNEP and BNEF, 2014).[5] The supply of electricity from wind turbines and solar photovoltaic (PV) panels recorded double-digit annual growth rates during this period. At the same time, technology costs have dramatically fallen in recent years – the price of crystalline silicon solar-PV cells has fallen by 80% since 2008, and by 99% since 1977 (The Economist, 2012).

However, new investment in clean-energy generation declined in 2012-13 – down to USD 256 billion in 2012, and to USD 232 billion in 2013. This decline has resulted primarily from policy uncertainty, which could continue to impede investment growth if not addressed. It also has resulted from the decreasing cost of solar-PV installations and the sector's consolidation following a situation of over-capacity (FS-UNEP and BNEF, 2014; BNEF, 2014a; Liebreich, 2014). Investment in clean energy rebounded in 2014, by 17% to USD 270 billion, largely driven by new investment in solar and wind energy (FS-UNEP and BNEF, 2015).

Clean-energy projects remain seriously constrained by specific investment barriers. Their risk-return profile can be less attractive than that of their fossil-fuel-based equivalents due to a range of factors, including market and government failures. Key policy obstacles include: inadequate carbon prices; inefficient fossil-fuel subsidies; a lack of a predictable and stable policy and regulatory environment (e.g. retroactive changes in support policies for clean-energy projects); market and regulatory rigidities that favour fossil-fuel incumbency in the electricity sector; a lack of investment in transmission, distribution and energy storage; the need to further integrate electricity markets; high financing costs; and outstanding barriers to international trade and investment (OECD, 2015; Baron, 2013). Governments can support private investment in clean energy, including through setting stable and predictable policy frameworks (OECD, 2015). This role is examined in the OECD *Policy Guidance for Investment in Clean Energy Infrastructure: Expanding access to clean energy for green growth and development* (OECD, 2015). Several OECD and non-OECD reports have also reviewed the impacts of trade and investment restrictions on international trade of clean-energy products and services (Bahar et al., 2013). Few reports, however, have considered the effects of such measures on international investment, which is the main purpose of the present report.

International investment accounts for an important share of clean-energy investment. Between 2004 and the first half (H1) of 2012, international investment has represented about one-third of asset finance[6] investment in clean energy (BNEF, 2013a). Asset finance of utility-scale projects accounted for the largest share (63%) of new investment in clean energy in 2014, though it has declined since 2011. Small-scale distributed capacity represented 27% (FS-UNEP and BNEF, 2015).

Trade and investment in solar PV and wind energy are increasingly interconnected within global value chains, especially in the solar-PV sector (OECD, 2013e; OECD, WTO and UNCTAD, 2014, 2013a). The prevalence of global value chains and rising international trade of intermediate goods in clean energy have implications for policy makers. Taking a value-chain approach can inform the design of effective policies to meet potentially conflicting goals, help reconcile policy trade-offs, and support the achievement of a level playing field for international investment in clean energy.

Figure 1.1 **Global asset finance investment in clean energy by origin of investor, 2004-H1 2012** (USD billion)

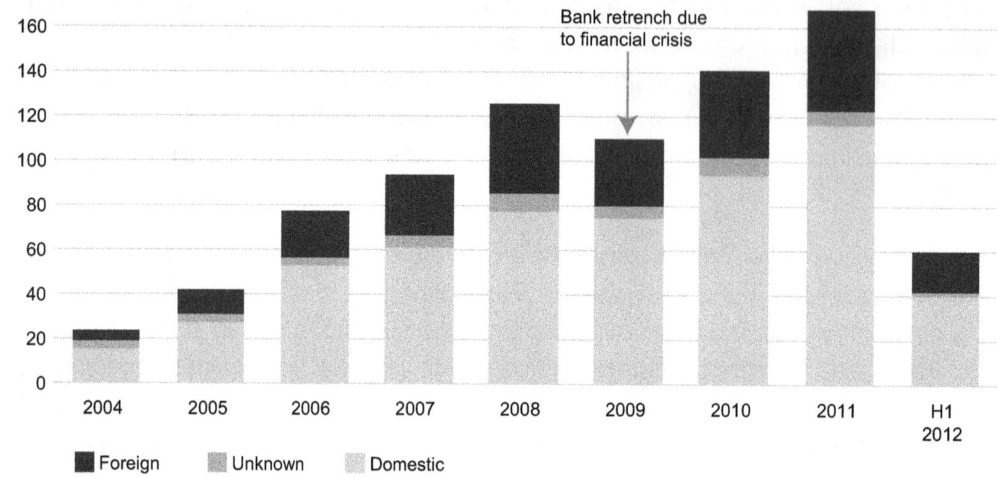

Source: BNEF (2013a).

This report considers solar PV and wind energy, and in particular midstream (manufacturing) and downstream activities (e.g. power plant installations, maintenance and operations).[7] Solar and wind power are the two dominant clean-energy sectors in terms of new investment flows (USD 114 billion and USD 80 billion respectively in 2013 and USD 150 billion and USD 100 billion in 2014;[8] Figure 1.2; FS-UNEP and BNEF, 2015). Other clean-energy sectors attract significantly smaller new investment flows.[9] Solar PV and wind energy are also expected to account for the largest shares of estimated investment needs to achieve the 2°C target under the IEA's 2DS scenario (29% and 23% respectively; IEA, 2012e). In addition, the solar-PV and wind-power sectors have experienced a recent increase of trade and investment restrictions such as LCRs, as discussed subsequently.

Figure 1.2 **G20 investment in clean energy by sector, 2004-13** (USD billion)

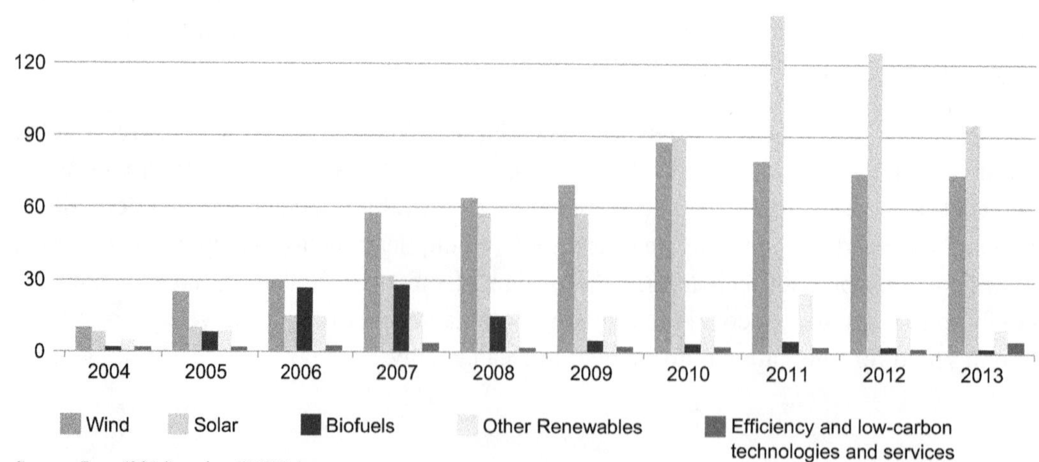

Source: Pew (2014), using BNEF data.

OECD countries and emerging economies account for the largest share of international (as well as domestic) investment in clean energy. From 2004 to the first half of 2012, North-North flows represented 77% average of international investment flows in clean energy (BNEF, 2012a). Emerging economies have had only a limited share of new international investment flows during the past few years. Overall however, total new investment (domestic plus international) has increasingly shifted towards emerging economies and developing countries, especially China (BNEF, 2012a, 2013c, 2014a; FS-UNEP and BNEF, 2014).

This report focuses on international investment in solar PV and wind energy in OECD countries and emerging economies, since they account for the largest share of inward and outward international investment in these sectors. It aims to assess policies that may hinder investment in the solar-PV and wind-energy sectors. The focus on solar and wind energy was chosen because these sectors have become the two dominant clean-energy sectors in terms of new investment flows (USD 150 billion and USD 100 billion respectively in 2015).

The role of policy support to scale up investment in clean energy

Recognising the role of clean energy in meeting climate change and economic growth-related objectives, policy makers have provided significant policy support to clean energy deployment over the past decade. Globally, subsidies for renewable energy amounted to USD 121 billion in 2013, which was 15% higher than in 2012 and included USD 82 billion to electricity generation from renewable energy and USD 19 billion to biofuels for transport (IEA, 2013c). By early 2014, 138 countries had implemented renewable-energy support policies, up from 127 countries in 2013 (IEA, 2014c; REN21, 2014). The most prevalent incentive policies are tax reductions (84 countries), feed-in tariffs[10] (FiT) and FiT premiums (71 countries), and auctions or tenders (45 countries) (IRENA, 2013a). Incentive policies such as FiTs, tax incentives and capital subsidies have played a significant role in driving renewable-energy investment and technology cost reductions (Bahar et al., 2013; Kitson et al., 2011). Such measures, however, are considered to be far less cost-effective in reducing GHG emissions than carbon pricing instruments, such as economy-wide emission trading systems or broad-based carbon taxes (OECD, 2013j). In the absence of an explicit price on carbon, or in addition to policies putting a price on carbon, many countries have implemented renewable-energy support policies, to address spillovers and bring down the cost of renewable energy. When such support policies are used, they need to be well-targeted and time-limited to help improve their relative cost-effectiveness.

Most of these forms of support typically have not established differentiated treatment between domestic and international investors. Several countries have also supported clean energy by removing trade and investment barriers for domestic and international investors in clean energy, such as import tariffs and restrictions to foreign direct investment (FDI).

Since the 2008 financial crisis, however, the perceived potential of clean energy to promote growth and employment has led several governments to implement green industrial policies favouring domestic solar-panel and wind-turbine manufacturers. In a post-crisis recovery context, such policies aimed to support domestic industries, create local jobs and value added and promote exports. Governments have notably set local-content requirements

(LCRs) in the solar-PV and wind-power industries. Local-content requirements (or "locally manufactured content requirements") for solar or wind energy typically require developers and investors, in order to be eligible for policy support or tenders, to source a specific share of manufactured components, equipment locally.

This report assesses the possible effects of such restrictions across the various segments of the solar-PV and wind-energy value chains,up to the end product: electricity.[11] It is often noted that measures such as LCRs, which support midstream manufacturers, can play an important role in building political support and public acceptance for renewable-energy incentive schemes. Factors other than the trade- and investment-restrictiveness of renewable-energy support policies may also have an even greater impact on international investment in solar PV and wind energy (e.g. policy uncertainty). These considerations notwithstanding, the increasing use of restrictions to international trade and investment warrants an examination of their impacts on downstream segments of the solar-PV and wind-energy value chains.

Key messages

By assessing the impacts of LCRs across different segments of the solar-PV and wind-energy value chains, this report provides policy makers with evidence-based analysis to guide their decisions in designing solar-PV and wind-energy support policies. The report finds that the impacts of LCRs, in terms of increasing domestic manufacturing and creating local jobs, are at best mixed. In countries that have a competitive advantage in solar- and wind-energy manufacturing, LCR policies directed at midstream manufacturing do not provide any additional positive impacts. Such countries should promote open international trade and investment in these sectors. The report suggests that policy makers seeking to expand renewable-power capacity in countries with a nascent or uncompetitive solar or wind-turbine industry, should address local impediments that hinder the manufacturing sector's competitiveness rather than impose requirements.

This report focuses on LCRs. Local-content requirements have been planned or implemented in solar- and wind-energy sectors in at least 21 countries, including 16 OECD countries and emerging economies, mostly since 2009. This has prompted five World Trade Organization (WTO) disputes since 2010. Other existing measures considered in this report that may hamper international investment in solar PV and wind energy include: technical barriers to trade; actionable subsidies; and FDI regulatory restrictions. Foreign Direct Investment restrictions in solar-PV and wind energy remain relatively limited in OECD countries. New research presented in this report, however, highlights the presence of divergent national standards[12] in wind energy and other non-tariff measures in selected OECD and emerging economies. In addition, the alleged use of dumping or actionable subsidies has resulted in an escalation in the use of trade remedies in solar-PV and wind energy. The WTO rules authorise countries to use trade remedies to address unfair trade practices and trade distortions. Since January 2010, developed countries and emerging economies have imposed nine anti-dumping duties (ADs) and seven countervailing duties (CVDs) on products associated with solar PV or wind energy, and launched 24 WTO investigations for ADs or CVDs.[13]

After taking stock of LCRs, the report assesses the possible effects of these requirements across the solar-PV and wind-energy value chains. Adopting a value-chain approach has implications for policy makers:

- Production and activities in the solar-PV and wind-energy sectors are increasingly global. In other words, domestic solar-PV and wind-power generation relies on an increasing share of imported intermediate products. Policies aimed at supporting midstream manufacturers, such as LCRs, are thus likely to hinder the profitability of downstream investors – e.g. by raising the cost of inputs, or reducing overall demand as costs are passed through to consumers.

- A value-chain approach also highlights the contribution of downstream activities to value added and investment. In the solar sector in particular, manufacturing activities represent only 18-24% of total jobs, according to recent estimates in the United States and worldwide. According to several studies, at least 50% of solar-PV jobs and value-added are located in downstream activities. Some of the benefits that policy makers expect from LCRs in midstream industries may thus be undermined by indirect effects on downstream segments of the value chains.

This report has not examined the extent to which countries' propensity to provide the support they have provided to date for wind-solar investment is sensitive to the use of LCRs. It provides empirical evidence that, other things being equal, LCRs may reduce the competitiveness of the wind and solar-PV sectors in countries that impose LCRs, divert investment to countries that do not impose LCRs, and reduce total international investment in the solar-PV and wind-energy sectors. In particular:

- Results from a new OECD econometric analysis[14] undertaken for the purpose of this report indicate that LCRs do not have positive impacts on international investment flows. While feed-in tariff (FiT) policies play an important role in attracting international investment in solar PV and wind energy, the analysis shows that, other things being equal, investment may flow to the country that does not combine support with a local-content requirement. The estimated detrimental effect of LCRs is even slightly stronger when both domestic and international investments are considered.

- According to results from the new 2014 OECD Investor Survey on "Achieving a Level Playing Field for International Investment in Clean Energy", conducted for this project, LCRs stood out as the main policy impediment for international investors in solar PV and wind energy.

- During an OECD roundtable consultation hosted in December 2013, private and public sector participants highlighted the need to differentiate between the short-term and long-term effects of LCRs. Policy makers often face policy trade-offs between short-term priorities (e.g. securing rapid increases in the creation of low-skill jobs) and long-term goals (e.g. achieving cost-effectiveness in clean-energy sectors and increasing the structural resilience of labour markets).

- Several recent country experiences with LCRs in solar and wind energy show that LCRs can raise the costs of downstream activities in the value chain, such as renewable-energy-based electricity generation, – because they mandate the use of higher-cost domestic inputs. Evidence suggests also that LCRs may not have been effective in several countries in generating domestic employment and value added across the solar- and wind-energy value chains, and that removing LCRs helps support technology transfer and innovation. Local-content requirements have also prevented cost reductions in domestic manufacturing activities and have had limited spillovers, especially in countries without sufficient domestic market size or local technical expertise.

Feed-in tariffs and capital subsidies are considered to be far less cost-effective in reducing GHG emissions than carbon pricing instruments (e.g. emission trading systems or broad-based carbon taxes; OECD, 2013r). However, to date, many more countries have opted to implement renewable-energy support policies (more than 100) rather than explicit carbon pricing instruments (approximately 40). This is because direct support to renewable energy can help achieve multiple policy goals, beyond GHG emissions reduction. Where renewable-energy support policies are used, they will need to be well-targeted and time-limited to help improve their relative cost-effectiveness.

Policy makers could usefully consider alternatives to LCRs to support their domestic solar-PV and wind-power industries. Policy options include: well-targeted support to research and development (R&D) and innovation in solar and wind-power technologies; and training programmes and promotion measures to build technological skills and local capability. Conversely, well-designed and predictable incentive measures (such as feed-in tariffs with no LCRs attached to them) or more effective carbon pricing instruments (such as carbon taxes and tradable permits with predictable and rising prices) can effectively support the deployment of solar and wind energy to mitigate climate change and build an industrial base without restricting international trade and investment.

The report also outlines the possible impacts of other measures such as technical barriers to trade based on new research undertaken in co-operation with the European Wind Energy Association (EWEA). In the wind-energy sector for instance, divergent national standards on wind-turbine designs directly impact trade flows, and can also increase transaction costs for international investors. The report also discusses the escalation of trade disputes related to trade remedies, which are used to address distortions and unfair trade practices but seem to have seriously impacted international trade flows. The total reduction of trade as a result of trade remedies is estimated to reach USD 8.5 billion for crystalline silicon PV cells alone. In the EU for instance, trade remedies in renewable energy represents almost 75% of the total amount of all the EU's trade remedies in force today.

Although this report recognises the reasons behind the use of LCRs, it advises against their implementation. Supporting open, competitive and demand-driven solar-PV and wind-energy sectors would help sustain the trend towards cost reductions and make renewable energy more competitive vis-à-vis fossil-fuel energy. This, in turn, would reduce the cost of policy support to renewable energy, while driving investment. Evidence-based analysis is needed to improve the coherence of renewable-energy support policies and reduce their cost.

International co-operation is needed to align trade and investment policy in clean energy, in a context of global value chains, including to respond to the escalation of local-content requirements and trade remedies or to improve international recognition or harmonisation of standards (e.g. in wind energy). In particular, countries interested to further international co-operation on climate change and green growth could agree on a new Sustainable Energy Trade Agreement (SETA), to address trade governance issues related to clean energy, as well as the resulting legal challenges and opportunities. The *Joint Statement Regarding Trade in Environmental Goods*, signed by 15 WTO member countries in January 2014, is an encouraging sign of governments' commitment to achieve global free trade in environmental goods. The signing countries announced that they were launching a new multilateral initiative to reduce tariffs on a selected list of environmental goods by 2015. International co-operation is also needed to improve the monitoring and measurement of subsidies, and to improve possible anti-competitive behaviour of state-owned enterprises (SOEs).

Improving tracking and monitoring of trade and investment restrictions in solar PV and wind energy, and in green sectors more broadly, would enable better analysis of their impacts. Monitoring of trade and investment measures across sectors or in the energy sector already exists. There is, however, no single source that tracks all trade and investment measures in clean energy.

More research and policy efforts are needed to assess the implications of trade and investment restrictions for both national and domestic investment, across green value chains and beyond the solar-PV and wind-energy sectors. Future work could usefully consider ways of sharing lessons learned and encouraging good practices in designing incentives aimed at supporting green industries without distorting international investment. In addition, further research is needed on the impact of investment restrictions on green services, especially in downstream segments of the value chains. Further work is also needed on distortive practices in government procurement of green infrastructure projects such as renewable-energy plants and transport infrastructure. Finally, further work is needed to assess operational obstacles to international investment in clean energy, such as grid access.

Structure of the remainder of the report

Chapter 2 of the report provides an overview of the solar-PV and wind-energy value chains. It highlights the rising share of imported intermediate goods in the value chains and the importance of downstream segments for generating value added and employment. Chapter 3 takes stock of the increased use of LCRs in solar PV and wind energy, and assesses the possible implications of such measures for international trade and investment within key segments of the solar- and wind-energy value chains. An annex to Chapter 3 provides evidence from an econometric analysis of the impact of local-content requirements on international investment flows. Another annex to Chapter 3 summarises the results from the 2014 OECD Investor Survey on "Achieving a Level Playing Field for International Investment in Clean Energy." Chapter 4 briefly takes stock of other policies that provide differentiated treatment between foreign and domestic investors in solar PV and wind energy, including divergent national standards in wind energy. An annex to Chapter 4 provides details on an analysis of technical restrictions to international investment in wind energy in Brazil, Canada, China, India, and South Africa; factsheets are provided for each country.

Notes

1. "Clean energy" as defined in this paper includes the following sectors: solar, wind, small and large hydroelectric, geothermal, marine, biomass and waste-to-energy power plants, carbon capture and storage (CCS) technologies and energy-efficient technologies such as smart grids and electric vehicles.

2. In the 2010 Cancun Agreements, Parties of the United Nations Framework Convention on Climate Change (UNFCCC) agreed to work together with a view to reducing global greenhouse gas emissions so as to hold the increase in global average temperature below 2 °C above pre-industrial levels.

3. In 2013, fossil-fuel consumption subsidies amounted to USD 548 billion globally while support for fossil-fuel production and consumption in OECD countries was about USD 55-90 billion per annum in recent years; IEA (2014c); OECD (2013b).

4. I.e. greenfield or new build investment.

5. Total investment in clean energy reached USD 378 billion in 2011, including: new investment in clean energy as defined by BNEF (USD 279 billion); project acquisitions and refinancing, corporate M&A, private equity buy-outs and public markets investor exits (USD 73 billion); and investment in large hydro (USD 26 billion); FS-UNEP and BNEF (2014, 2012).

6. As defined by BNEF (2012a), asset finance for investment in renewable energy includes electricity generation and biofuels production assets that meet the following size criteria: 1 MW or larger for biomass and waste, geothermal, solar and wind-energy generation; 1-50 MW for hydroelectric projects, any size for marine-energy projects, and 1million litres per year or greater for biofuel projects. The financing of carbon capture and storage and energy-smart technologies, along with M&A and refinancing deals are excluded.

7. The report does not consider upstream activities linked to the production and extraction of raw materials and minerals.

8. Representing respectively 56% and 37% of new investment in renewable energy generation and fuels in 2014, without considering hydro power larger than 50 MW; BNEF (2013a); FS-UNEP and BNEF (2015).

9. Including: smart technologies, e.g. smart grids, energy efficiency and electric vehicles (USD 35 billion in 2014); biomass and waste-to-energy (USD 8 billion); hydroelectric power (including USD 5 billion in small hydro and about USD 35 billion in large hydro); biofuels (USD 5 billion); and geothermal power (USD 3 billion); FS-UNEP and BNEF (2015).

10. Feed-in tariffs and premiums (or bonuses) are price-driven incentives for the production of electricity from renewable energy sources, and are typically differentiated by technology type and size of installation.

11. A value chain consists of the full range of activities in a given sector that firms undertake to bring a product or service from its conception to its end use by final consumers; OECD (2013d).

12. I.e. national standards that differ from international standards.

13. This report does not consider trade disputes linked to upstream production of raw materials; updated as of September 2014.

14. This econometric analysis was conducted and authored by Miguel Cárdenas-Rodríguez to measure the impact of LCRs on global international investment flows in solar-PV and wind-energy generation from 2000 until 2011.

References

Bahar, H., J. Egeland and R. Steenblik (2013) "Domestic incentive measures with possible trade implications", *OECD Trade and Environment Working Papers*, No. 2013/01, OECD Publishing, Paris. http://dx.doi.org/10.1787/5k44srlksr6f-en.

Baron, R. (2013), "Renewable energy: a route to decarbonisation in peril?", *the 29th Round Table on Sustainable Development*, OECD, Paris.

BNEF (2014a), "Clean energy investment falls for second year", Press Release of Bloomberg New Energy Finance, London.

BNEF (2013a), "New investment in clean energy fell 11% in 2012", Press Release of Bloomberg New Energy Finance.

BNEF (2012), "North-south clean energy investment flows: An $8bn step to a $100bn goal", Clean Energy White Paper of Bloomberg New Energy Finance. Economist (The) (2012), "Sunny uplands: Alternative energy will no longer be alternative", in *The World in 2013*, print edition, The Economist, London.

Frankfurt School (FS)-UNEP Centre and BNEF (2015), *Global Trends in Renewable Energy Investment 2015*.

Frankfurt School-UNEP Centre and BNEF (2014), *Global Trends in Renewable Energy Investment 2014*, Frankfurt School of Finance & Management gGmbH.

IEA (2014a), *Energy Technology Perspectives 2014*, IEA/OECD Publishing, Paris. http://dx.doi.org/10.1787/energy_tech-2014-en.

IEA (2014b), *Policy and Measures Databases* (database), www.iea.org/policiesandmeasures (accessed 27 August 2014).

IEA (2014c), *World Energy Outlook 2014*, IEA/OECD Publishing, Paris. http://dx.doi.org/10.1787/weo-2014-en.

IEA (2013a), *Tracking Clean Energy Progress, IEA Input to the Clean Energy Ministerial*, OECD/IEA Publishing. Paris. http://www.iea.org/publications/tcep_web.pdf.

IEA (2013b), *PVPS Report: A Snapshot of Global PV 1992-2012*, Report IEA-PVPS T1-22:2013, OECD/IEA Publishing, Paris.

IEA (2013c), *World Energy Outlook 2013*, IEA/OECD Publishing, Paris. http://dx.doi.org/10.1787/weo-2013-en.

IEA (2012a), *Tracking Clean Energy Progress: Energy Technology Perspectives 2012 excerpt as IEA input to the Clean Energy Ministerial*, OECD/IEA Publishing, Paris. www.iea.org/media/etp/Tracking_Clean_Energy_Progress.pdf.

IEA (2012b), *Energy Policies Beyond IEA Countries: Ukraine 2012*, OECD Publishing. Paris, http://dx.doi.org/10.1787/9789264187450-en.

IEA (2012c), *World Energy Outlook 2012*, OECD/IEA Publishing, Paris. http://dx.doi.org/10.1787/weo-2012-en.

IEA (2012d), *Energy Technology Perspectives 2012: Pathways to a Clean Energy System*, IEA/OECD Publishing, Paris. http://dx.doi.org/10.1787/energy_tech-2012-en".

IRENA (2013a), *Renewable Energy and Jobs* International Renewable Energy Agency, Abu Dhabi.

IRENA (2013b), *Renewable Energy Auctions in Developing Countries*, International Renewable Energy Agency, Abu Dhabi.

Liebreich, M. (2015), Presentation from Bloomberg New Energy Finance Summit, New York, 14 April 2015, http://about.bnef.com/content/uploads/sites/4/2015/04/BNEF_2014-04-08-ML-Summit-Keynote_Final.pdf.

Liebreich, M. (2014), Presentation from Bloomberg New Energy Finance Summit, New York, 7 April 2014.

OECD (2015), *Policy Guidance for Investment in Clean Energy Infrastructure: Expanding access to clean energy for green growth and development*, OECD Publishing, Paris. http://dx.doi.org/10.1787/9789264212664-en.

OECD (2013a), *OECD Code of Liberalisation of Capital Movements,* OECD Publishing, Paris, http://www.oecd.org/daf/inv/investment-policy/CapitalMovements_WebEnglish.pdf

OECD (2013b), *Inventory of Estimated Budgetary Support and Tax Expenditures for Fossil Fuels 2013*, OECD Publishing, Paris. http://dx.doi.org/10.1787/9789264187610-en.

OECD (2013c), *Interconnected Economies: Benefiting from Global Value Chains*, OECD Publishing, Paris. http://dx.doi.org/10.1787/9789264189560-en.

OECD (2013d), *OECD Glossary of Statistical Terms*, http://stats.oecd.org/glossary/.

OECD (2013e), "OECD-WTO Database on Trade in Value-Added FAQs: Background Note", OECD, http://www.oecd.org/sti/ind/TIVA_FAQ_Final.pdf.

OECD (2013f), "The Climate Challenge: Achieving Zero Emissions", lecture by the OECD Secretary-General, Mr. Angel Gurría, London, 9 October 2013.

OECD (2013g), "OECD roundtable consultation on achieving a level playing field for international investment in solar and wind energy", Paris, 6 December 2013, Summary report.

OECD (2013h), *Maintaining a Level Playing Field between Public and Private Business for Growth and Development: Background Report*, Meeting of the OECD Council at Ministerial Level, 29-30 May, 2013, OECD Publishing, Paris.

OECD (2013i), "How to unlock private investment in support of green growth?", OECD Green Growth and Sustainable Development (GG-SD) Forum, Paris, 5-6 December 2013.

OECD (2013j), *Effective Carbon Prices,* OECD Publishing, Paris.

OECD, WTO and UNCTAD (2014), "Report on G20 trade and investment measures (mid-November 2013 to mid-May 2014)".

OECD, WTO and UNCTAD (2013a) "Implications of global value chains for trade, investment, development and jobs", prepared for the G-20 Leaders Summit at Saint Petersburg, September 2013.

OECD, WTO and UNCTAD (2013b), "Report on G20 trade and investment measures (mid-October 2012 to mid-May 2013)".

REN21 (2014), *Renewable 2014 Global Status Report,* Renewable Energy Policy network for the 21st Century.

Chapter 2

Key trends within the solar- and wind-energy global value chains

This chapter provides an overview of the solar-PV and wind-energy global value chains, discussing each in turn. It describes how international trade and foreign direct investment in greenfield projects have contributed to the emergence of global value chains in both solar-PV and wind-energy sectors. For both sectors, the report discusses trends in production, trade volumes, industry concentration, as well as domestic and international investment. Similarities and differences between the two sectors are noted. Particular attention is given to the increasing reliance of domestic production on imported intermediate inputs. The chapter highlights the growing importance of downstream activities (project development, installation, and maintenance) compared to midstream activities (manufacturing) in terms of value added, employment and investment, and discusses some of the policy implications.

Brief overview of the solar-PV and wind-energy value chains

Solar PV

The solar-PV sector can be broken down into three main segments: upstream, midstream and downstream activities (Figure 2.1). Crystalline silicon (c-Si) PV technology accounted for more than 89% of the global market for PV cells and modules in 2011, while thin-film PV accounted for almost all of the remainder (IEA and IRENA, 2013). In the case of c-Si technologies, upstream activities start with polysilicon production. Midstream activities melt polysilicon into ingots, which are subsequently cut into wafers and then processed into solar cells and ultimately modules. Downstream activities include power-plant project development, system integration, plant installation, maintenance and other related services.

Figure 2.1 **Overview of the solar-PV energy value chain** (c-Si technology)

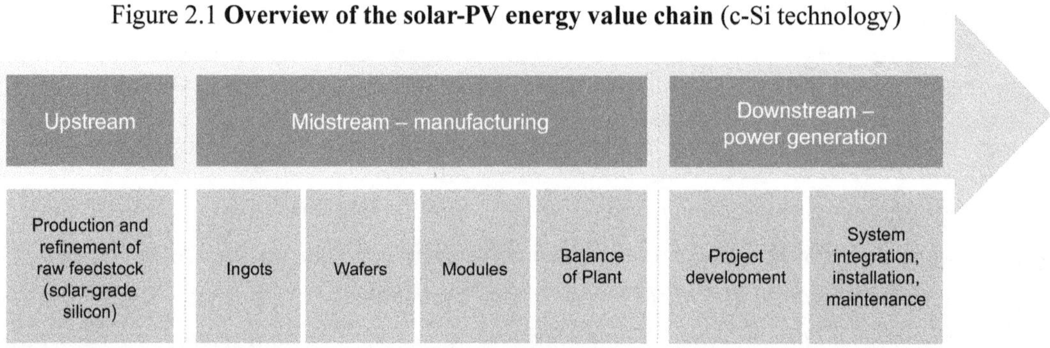

Wind energy

The wind-energy value chain (onshore and offshore) can be broken down into two main segments. First, upstream and midstream manufacturing activities include the design, production and assembly of wind turbines and components such as the gearbox and generator (housed in the "nacelle"). Second, downstream activities and services relate to: the deployment and utilisation of wind electric power, including project and pre-project financing, project development, and post-project operations and maintenance; engineering, procurement and construction of the power plant; and grid connection and power sales (Figure 2.2; Lema et al., 2012, 2011). The offshore wind-energy value chain also includes the construction and installation of substations, foundations and cables.

Figure 2.2 **Overview of onshore wind-energy value chain**

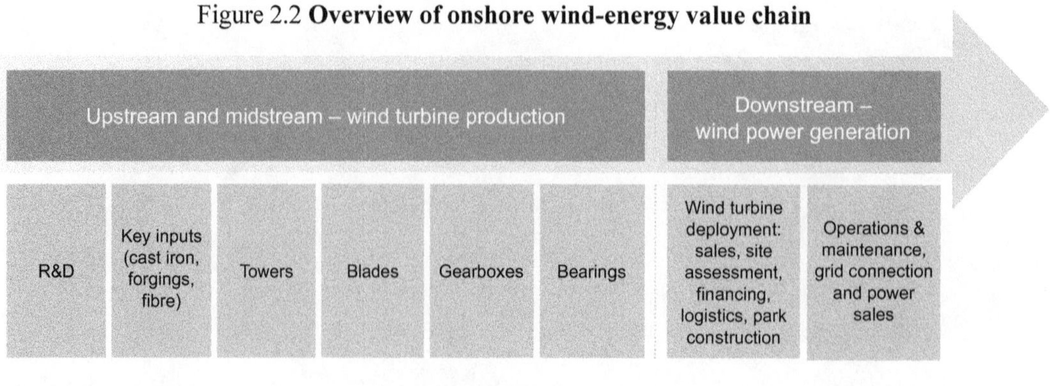

Key trends within the solar-PV and wind-energy value chains: implications for policy makers

The role of international trade and FDI in driving the global integration of the value chains

International trade and foreign direct investment (FDI) in greenfield projects have strongly contributed to the growth and global integration of the solar-PV and wind-energy sectors. Both industries – and especially solar PV energy – are characterised by the emergence of global production networks. These networks allow firms to optimise production costs by sourcing activities such as equipment manufacturing based on countries' comparative advantages in terms of costs, skills and materials and other location-specific factors (OECD, WTO and UNCTAD, 2014, 2013a; OECD, 2013e; ICTSD, 2011a).

Solar PV

International trade in solar-PV equipment has increased considerably in the past decade, in line with installed solar-PV capacity (in GW). Trade volumes increased rapidly from 2006 until 2011 among leading solar-PV markets. There are significant levels of "two-way trade", as top PV markets are both major importers and exporters of solar-PV manufactured goods (Tables 2.1, 2.2):

- In the United States for instance, total solar-PV imports increased from USD 434 million in 2006 to USD 2.6 billion in 2010. At the same time, solar-PV US exports increased from USD 443 million in 2006 to USD 1.4 billion in 2010 (USITC, 2011). China accounted for 56% of US total solar-PV cell and module imports in 2011 (Platzer, 2012; USITC, 2011);
- Chinese PV exports increased by a factor of 26 between 2000 and 2010. During the same time, imports to China increased by a factor of 12 (Cao and Groba, 2013).

Table 2.1 **Imports of solar-PV cells, other photosensitive semi-conductor devices and LEDs, 2004-12** (USD millions)

	2004	2005	2006	2007	2008	2009	2010	2011	2012	CAGR** 2004/11 (%)	CHANGE 2011/12 (%)
EU *	2953	4096	5534	8411	17129	15157	30640	27416	13857	37	-49
China	1931	2362	2681	3289	3744	3607	6145	6720	6433	20	-4
United States	1251	1391	1848	2156	2760	2592	4412	7193	7260	28	1
Japan	1002	1136	1207	1131	1412	1212	2189	2306	3100	13	34

Notes: * Excluding intra-EU trade; ** Compound annual growth rates.
Source: UNEP (2014), using UN International Trade Statistics (Comtrade) database.

Table 2.2 **Exports of solar-PV cells, other photosensitive semi-conductor devices and LEDs, 2004-12** (USD millions)

	2004	2005	2006	2007	2008	2009	2010	2011	2012	CAGR** 2004/11 (%)	CHANGE 2011/12 (%)
EU *	689	764	1073	1260	2025	1748	1835	2203	2102	18	-5
China	644	1258	2460	5252	11745	10721	25179	27946	17483	71	-37
United States	1193	1298	1298	1582	1976	2018	2706	2427	1804	11	-26
Japan	4629	4796	5199	5472	6190	4673	6397	6604	5835	5	-12

Notes: * Excluding intra-EU trade; ** Compound annual growth rates.
Source: UNEP (2014), using UN International Trade Statistics (Comtrade) database.

Largely driven by international trade, the solar-PV energy sector has shifted from a European-dominated market into a global one. China, Japan and the United States became the top three solar-PV installers in 2013 (in GW), followed by Germany (REN21, 2014, 2013; Figure 2.3). In the manufacturing segment, China has become the largest global exporter of solar-PV manufactured goods. The assembling of solar panels now includes components and services produced in numerous countries. For example:

- Asian companies accounted for 87% of PV module production in 2013 (REN21, 2014);
- Germany (e.g. with the firm SMA Solar Technology), Switzerland (ABB), Japan (Omron, TMEIC and Tabushi) and China (Sungrow) rank among the largest producers of solar inverters[1];
- China and India are among the largest producers and exporters respectively of stainless steel and steel used in the construction and framing of solar panels (AFASE, 2012); and
- The United States (Hemlock Semiconductor), Germany (Wacker Chemie) and China (LDK Solar, GCL and Daqo) rank among the top producers and exporters of polysilicon.

The fabrication of solar-PV cells and modules is relatively concentrated, although to a lesser extent than wind-turbine manufacturing. The top 15 PV module manufacturers (9 of which are now Chinese) account for 50% of global production. The PV market is undergoing consolidation, due to a situation of excess manufacturing capacity and policy uncertainty since 2012.

The solar-PV energy sector has shifted from a European-dominated market into a global one. China, United States, Germany and Spain became the four leading markets for solar-PV power-generating capacity as of 2013 in response to the incentives provided by domestic support measures (REN21, 2014). By 2013, Asia accounted for 87% of module production globally, with China accounting for 67% of total production worldwide – in part thanks to the strong support provided under China's Five-Year Plan (IEA, 2013b; REN21, 2014).

A large increase in module production over the past few years, and a significant slowdown of investment in installed generating capacity in 2012 resulted in overcapacity in the market for manufactured solar-PV components. The manufacturing capacity for polysilicon, wafers and c-Si cells and modules is expected to exceed the projected demand for solar-

PV installations until at least 2016 (GTM, 2012). In response to this overcapacity, market consolidation and vertical integration have been underway to achieve cost rationalisation and economies of scale[2] (Wesoff, 2013). Price reductions, increased competition among manufacturers, considerable decreases in equipment prices and technology cost reductions have occurred in recent years. For example, prices of silicon solar modules fell by 30% in 2012 alone and by 80% since 2008 (REN21, 2013; Ghosh et al., 2012). Although solar-PV module prices stabilised in 2013, global investment in solar energy in 2013 was 20% lower than in 2012, down to USD 113.7 billion (FS-UNEP and BNEF, 2014; REN21, 2014). Along with other factors, the decline in investment stems from investors' uncertainty regarding regulatory support to solar PV (and wind) energy (REN21, 2013). Investment in solar energy rebounded in 2014 to USD 150 billion, largely driven by new utility-scale investment in China and in small-scale projects in Japan (FS-UNEP and BNEF, 2015).

Figure 2.3 **Market shares of top 15 solar-PV module manufacturers, 2012**

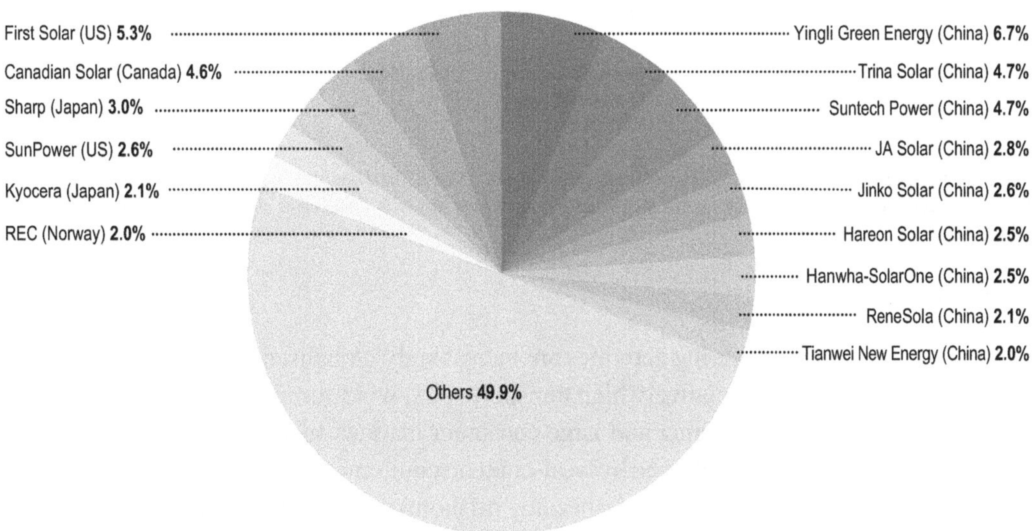

Sources: REN21 (2013), quoting GTM (2013); based on 35.5 GW produced in 2012.

In terms of market structure, the fabrication of solar-PV cells and modules is relatively concentrated, although to a lesser extent than wind-turbine manufacturing (Goodrich et al., 2011; Kirkegaard et al., 2010). The top 15 PV module manufacturers (nine of which are now Chinese) account for 50% of global production (REN21, 2013; Figure 2.3). Downstream solar-PV activities, by contrast, are local and fragmented. Recently however, new entrants have increased competition and cross-border flows. Upstream silicon production is highly concentrated in the EU, Japan and the United States where market incumbents retain a competitive advantage because of high entry costs due to the capital intensity of polysilicon production and the incumbents' use of advanced technology.

Greenfield FDI has been a significant driver of solar-PV deployment, although to a lesser extent than for wind energy. In 2012, around one-third of total asset finance[3] in solar power came from international investment (BNEF, 2013c, 2012). Conversely, brownfield FDI – through cross-border merger and acquisition (M&A) – remains smaller and limited to downstream activities.[4]

Figure 2.4 **Greenfield FDI (and cross-border acquisitions) in the solar sector (number of projects)***

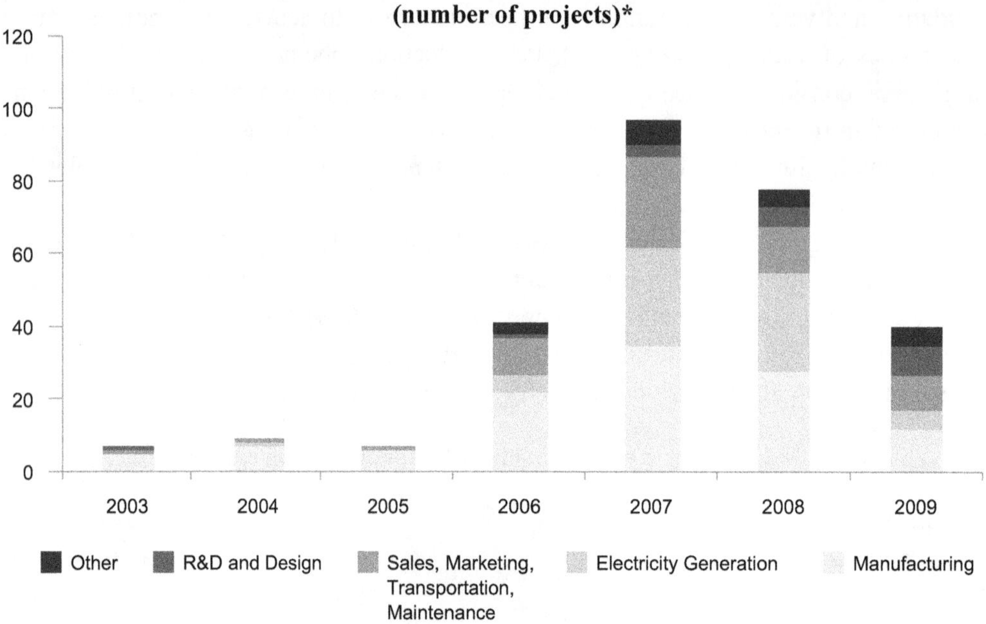

Source: Kirkegaard et al. (2010), using fDi Intelligence and Thomson ONE databases;
*All solar sector and related sub-sectors; includes cross-border investment with final stakes classified as direct investment; OECD (2008b).

Wind energy

Wind-turbine manufacturing activities are increasingly globalised and concentrated. Due to intense competition and relatively high transport costs, wind-turbine manufacturers tend to invest in plants close to emerging and large consumer markets such as Brazil, India and the United States. This trend is reinforced by local-content requirements (as discussed in Chapter 3). In addition, the wind-power industry (especially offshore) is undergoing a vertical-integration process, as manufacturers are trying to expand into downstream activities in order to diversify risk and secure their margins (REN21, 2013). As a result, the wind equipment manufacturing segment is highly concentrated, with the world's top ten turbine manufacturers accounting for 69.5% of the market in 2013, located primarily in China, Europe, India and the United States (Figure 2.5). By contrast, downstream activities such as system integration and installation typically remain local and fragmented (BNEF, 2013; Lema et al., 2012, 2011).

Greenfield FDI has become the dominant mode of global integration of the wind-energy industry, especially in the EU and the United States. While the growth in China's market still stems primarily from domestic investment, greenfield FDI has been a key driver of growth of US and European wind power generation markets (BNEF, 2012; Krüger, 2011). Total FDI and domestic investment in wind power generation grew steadily between 2004 and 2010, reaching USD 96 billion in 2010 (before declining in 2011-12) mostly located in Asia and Oceania (43%) and the Americas (29%; IEA, 2013a; BNEF, 2013). In the manufacturing segment, FDI inflows in host countries are increasingly spread across developed and emerging markets (Krüger, 2011). Compared with greenfield FDI, the share of M&As in total investment in wind electric power remains limited (26% in 2012), and has been declining since the financial crisis.

International trade has played a relatively less important role in driving the integration of the wind-power industry than in the case of the solar-PV sector. This is mostly due to high transport-related costs for wind-turbine equipment, which is much larger in size than for solar-PV components, and more difficult to transport since several wind turbine parts require special transportation equipment (Helpman, Melitz and Yeaple, 2004; Kirkegaard et al., 2009). It is also a result of technical barriers such as divergent national standards (Chapter 4). The level of cross-border trade intensity in the wind industry has even declined since 2006. This is notably related to the increase of greenfield FDI by top wind manufacturers in emerging markets, such as China. As a result, the share of "local content" (i.e. that is produced locally by domestic and foreign-owned firms) has increased, with regional hubs supplying key consumer markets. This helps explain why tariffs remain relatively low in the wind industry, unlike non-tariff measures such as local-content requirements (Chapter 3).

Largely driven by greenfield FDI, wind energy has become the second-largest clean-energy technology after large hydro power in terms of total installed capacity, led by China, Germany and the United States (Pew, 2013; REN21, 2013). Wind power capacity has been growing exponentially in the past decade, increasing at a compound rate of 25% since 2004, reaching 318 GW in 2013 (REN21, 2014). This rapid growth has led to price reductions, mainly due to economies of scale and excess capacity of turbine manufacturing plants. By 2012, prices had fallen by 20-35% from their 2008 peak (REN21, 2013). This trend benefitted downstream activities by increasing demand for wind projects. However, over-capacity and declining prices, combined with reductions in public support, have led to market consolidation and concentration (REN21, 2013).

Figure 2.5 **Market shares of top 10 wind turbine manufacturers, 2013**

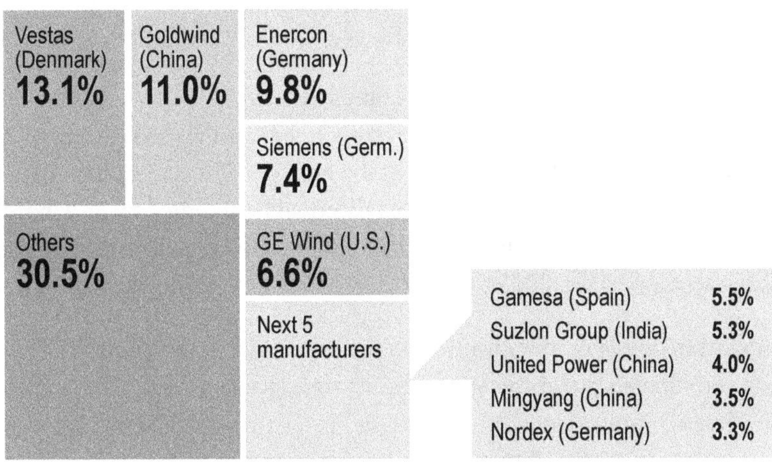

Source: REN21 (2014), based on data from Navigant Research; based on total sales of around 37.5 GW.

The increasing share of imports of intermediate goods in solar PV and wind energy

The rise of global value chains (GVCs) in the solar-PV and wind-energy sectors has implications for policy makers. It means that a country's domestic production and exports rely on an increasing share of intermediate goods imported from other countries (OECD, WTO and UNCTAD, 2013a). In the solar-PV sector in particular, while local companies account for a large share of the value added of downstream activities (as discussed subsequently), foreign companies account for an increasing share of value added for solar-PV module manufacturing. Solar-PV trade intensity[5] – estimated to be around 60-90% in 2006-08 – is significantly higher than in other clean-energy sectors (Kirkegaard et al., 2010). The share of intermediate PV imports and growth in global value chain activity – as measured by the OECD-WTO Trade in Value Added (TiVA)[6] initiative – has considerably increased in recent years, especially in the PV module manufacturing segment:

- In the US, a large share of PV inputs are imported: 27% of the total PV system value came from foreign sources in 2010, versus 70% for the value added of PV modules (GTM, 2011);
- In the EU, around 30% of the total gross value is generated outside the EU. In the EU market for PV modules however, Chinese manufacturers have captured more than 80% of the market share (Curtin, 2013; Ehrentraut O. et al., 2013).

Policy implications

The increasing share of PV imports can help explain why governments have increasingly used local-content requirements, to limit PV imports and favour domestic manufacturers (as discussed in Chapter 3). Yet, the emergence of global production networks and the rise of multinational enterprises (MNEs) have policy implications (OECD, 2013e). The expansion of international trade in intermediate products (including intra-firm) means that:

- Policies aimed at supporting manufacturing companies in midstream segments of the value chain, such as local-content requirements, are likely to hinder the profitability of downstream investors. This is because LCRs can increase the cost of inputs or reduce overall demand as costs are passed through to the market (as discussed subsequently in Chapter 3 of the report). Such measures may even ultimately hinder domestic manufacturing by impeding cost reduction in the sector (Barker, 2013; ICTSD, 2011a; REN21, 2013; Wu and Salzman, 2014).
- The definition of origin is more complicated to assess within the context of trade remedy investigations conducted under the WTO framework (Kasteng, 2014). As emphasised by former EU Trade Commissioner, Karel de Gucht, "[i]t is a fact that a lot of our imports are inputs for manufacturing that takes place here and that a significant share of the value of the finished goods we import has its origin in Europe" (De Gucht, 2012). The difference between "Made in" and "Made by" has implications for the relevance of trade measures, such as trade remedies, as discussed briefly in Chapter 4. It might increase the use of the rules of origin[7] for specific purposes. Countries can notably amend the rules of origin for specific products for facilitating the establishment of origin and the subsequent imposition of trade remedies on these products. The WTO, however, prohibits the use of rules of origin as a trade policy instrument (Kasteng, 2014).

The importance of downstream employment and value added

In addition to manufacturing activities linked to the production and assembly of components, downstream activities such as construction, installation, system integration, operations, maintenance and sales are employment-intensive (Goossens, 2012; CEM, 2011; Smith, 2014). The share of downstream employment and value added is particularly large in the solar sector. Although solar-PV manufacturing has the potential to generate jobs and create value, especially in emerging markets, the majority of jobs in the solar value chain lie in downstream activities (NRDC, 2012). In mature markets, manufacturing activities represented only 18-24% of total jobs across the solar value chain according to recent estimates in the United States and worldwide (Table 2.3; NRDC, 2012; Hufbauer et al., 2013). According to several studies, at least 50% of solar-PV jobs and value added are located in downstream activities (Table 2.3 and Figure 2.6).

Table 2.3 **Distribution of jobs across the solar value chain**

Data source	Manufacturing	Installation	Sales	Other
EPIA and Greenpeace[1] (global average for 2010 to 2020)	**25-40%**	**60-70%**		**Up to 5%**
Institute for Sustainable Futures[2] (global average for 2008)	**24%**		**76%**	
Solar Foundation[3] (US average for 2010 to 2011)	**15%**	**55%**	**26%**	**4%**

1. EPIA and Greenpeace (2006, 2011); NRDC analysis is based on the number of jobs expected to be created globally across the solar value chain in 2010 and 2020.

2. Rutovitz and Atherton (2009).

3. The Solar Foundation reports the number of US jobs in 2010 and 2011 (approximately 100 000 total). Numbers may not reflect different sectors' predominance in the US and the import-export imbalance; The Solar Foundation (2011).

Source: NRDC (2012).

Figure 2.6 **More than half of generated value lies downstream of module production**

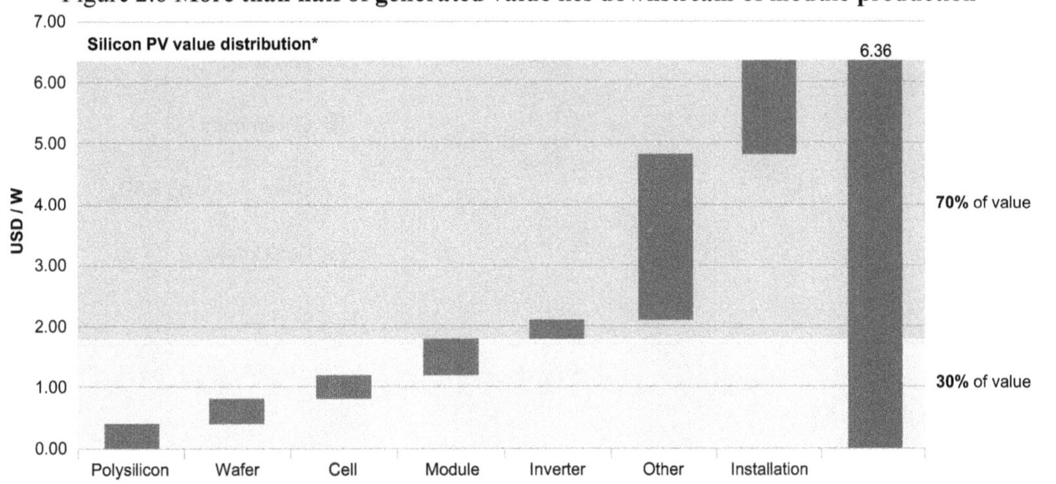

Based on unsubsidised value chain analysis of US silicon PV market. Roughly similar value distribution for thin-film technologies.

Source: NRDC (2012); quoting GTM Research (2011); European Photovoltaic Industry Association (EPIA) and Greenpeace (2006, 2011); Rutovitz and Atherton (2009); The Solar Foundation (2011).

In the US silicon PV value chain, for instance, cells and modules manufacturing account respectively for 20% and 10% of jobs, while the remaining 70% lie in downstream activities (Box 2.1; NRDC, 2012). In the US thin-film solar value chain, modules represent about 40% of the value, while the remaining 60% lie in downstream activities).

Box 2.1 **The job intensity of the different segments of the US solar-PV value chain**

Downstream activities represent the largest share of jobs in the US solar sector and have the highest growth rate, according to the National Solar Jobs Census 2013.[1] The installation segment is the largest segment of the US solar industry in terms of employment in the sector (i.e. 49%, with around 70 000 solar workers out of 140 000; Figure 2.7 and Table 2.4). Employment in the installation sector has grown by nearly 60% over 2010-13. Conversely, manufacturing and assembling only accounts for 10% of jobs in the US solar sector, as a large share of the activities of US manufacturing firms is conducted abroad. Around 26% of solar manufacturing establishments in the United States serve as US headquarters for manufacturing that is conducted abroad, according to the 2013 survey.

Figure 2.7 **Percentage of respondents, by sector**

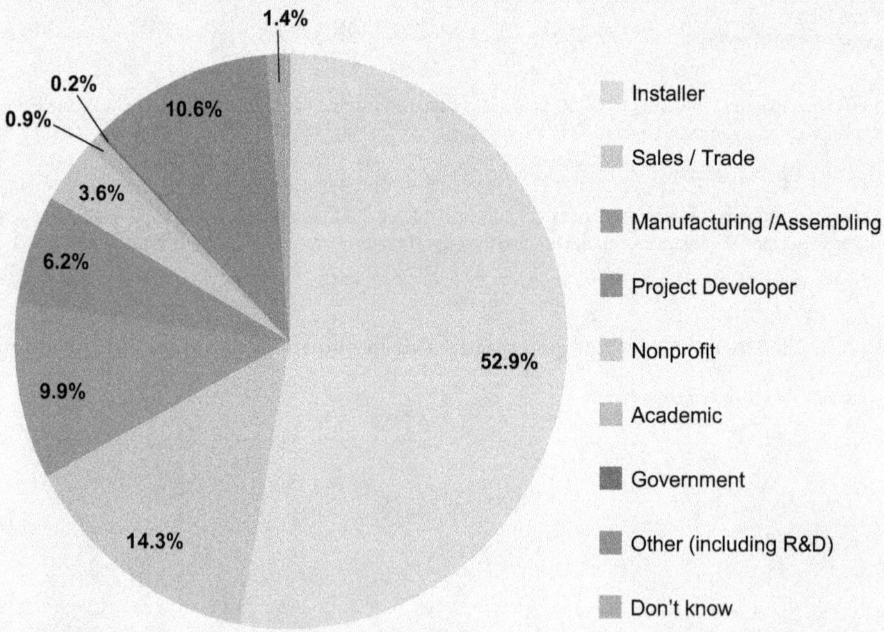

Box 2.1 **The job intensity of the different segments of the US solar-PV value chain**
continued..

Table 2.4 **Data by sector – number of solar workers** (in thousands)

Sector	2010 Jobs	2011 Jobs	2012 Jobs	2013 Jobs	2012-13 Growth rate	2014 Projected employment	2013-2014 Expected growth rate
Downstream	56	62	81	102	33%	119	13%
Installation	44	49	57	70	22%	84	21%
Manufacturing	25	38	30	30	0.4%	32	9%
Other	13	5	8	11	39%	13	16%
Total	94	105	119	143	20%	164	16%

[1] I.e. the fourth annual update of current employment and projected growth in the US solar industry, conducted by the Solar Foundation.

Source: The Solar Foundation (2014), "National Solar Jobs Census 2013: The Annual Review of the U.S. Solar Workforce".

Several studies estimate that manufacturing accounts for a greater share of employment in the wind-energy sector – around 30 to 60% of jobs – although there is less information available than for solar-PV energy. In the US, approximately 30% of total jobs in the wind-energy industry were in manufacturing activities as of 2012 (25 500 out of 80 700; AWEA, 2012). In the EU, 57% of direct employment in wind energy occurred in wind turbine and component manufacturing (77 564 out of 135 863). The remainder of jobs were associated with project development and services (EWEA, 2012).

Since 2012, domestic value creation remains significant in downstream activities, whereas midstream manufacturing is going through a consolidation process, as explained previously (REN21, 2014). Since 2008, the solar-PV and wind manufacturing sectors have experienced job losses in developed countries, especially in European countries. Conversely, the number of downstream installers has increased in the solar-PV sector during this time. Similarly, in the United States, most of the new jobs created in solar-PV energy since 2010 are in downstream activities (REN21, 2013).

Policy implications

Since a large share of the value-added and job-creation potential lie in downstream activities, especially in the solar-PV sector, policies targeting upstream and midstream manufacturing activities (such as local-content requirements) may not be effective in creating domestic jobs and value across the entire domestic supply chain. Indeed, local-content requirements, by favouring midstream manufacturers, may increase the cost of intermediate inputs for downstream installations and construction activities, increase the retail price of solar or wind electricity and create barriers for supply-chain optimisation. Given the relative importance of downstream activities compared with manufacturing activities, measures such as LCRs may have a limited or even negative impact in supporting investment, value creation and employment throughout the value chain as a whole (Chapter 3).

Conversely, research suggests that policy support to downstream activities may be more lucrative in terms of jobs and value creation, even in the short term (NRDC, 2012). Chapter 3 discusses the impacts of LCRs in solar and wind energy.

The share of downstream investment

In renewable energy, the share of investment in manufacturing is relatively small compared with the more capital-intensive downstream activities. Investment in power plants and other infrastructure assets represents the bulk of total investment. Globally, manufacturing equipment represented only 6% of new investment in renewable energy in 2013 (i.e. USD 12 billion out of USD 214 billion), including USD 1.4 billion from private equity expansion capital and USD 11 billion from public markets (FS-UNEP and BNEF, 2014).

In the solar sector, the share of greenfield FDI attributable to downstream activities, including electricity generation, sales and maintenance services, has considerably increased since 2006 (Figure 2.4). Manufacturers are also increasingly pushed to invest downstream due to vertical integration and industry consolidation. An increasing number of greenfield FDI and brownfield M&A transactions involving global firms are to be expected throughout the solar-energy value chain (Kirkegaard et al., 2010).

Box. 2.2 **Employment in solar and wind energy**

Available information on employment in renewable-energy sectors has considerably improved in recent years. An increasing number of sector-specific studies indicate that renewable-energy sectors can create new jobs (in gross terms). However, important gaps remain. There is a lack of reliable statistics and studies on the economy-wide net impact of renewable energy on employment. This report is not assessing the job potential of solar- and wind-energy sectors as a whole, but only the relative importance of midstream versus downstream activities in terms of employment intensity.

Globally, an estimated 6.5 million[1] people work directly or indirectly[2] in renewable-energy sectors (estimates based mostly on the 2012-13 period). The solar-PV sector has become the largest source of employment in the broader renewable-energy sector (2.3 million indirect or direct new jobs, in gross terms), followed by biofuels (1.5 million). The bulk of employment for the solar-PV industry is now concentrated in China (1.6 million), followed by India, the US and the EU (especially Germany). The estimate for China is substantially higher than estimates from previous years (in the 300 000–500 000 range), due to the massive and recent expansion of downstream installation jobs in China. The wind-power sector has created around 0.8 million new jobs (in gross terms), despite a recent slowdown in employment creation, due in part to policy changes.

Sources: REN21 (2014, 2013); IRENA (2013a); OECD (2013e); CEM (2011); Deloitte/AEE (2011); fDiIntelligence (2013).

In terms of policy implications, the importance of downstream investment means that local-content requirements, if they increase input costs for downstream activities, may hamper the attractiveness of solar and wind energy vis-à-vis fossil fuels for international investors, thereby slowing the flow of FDI to the solar-PV and wind-energy sectors.

Notes

1. In USD total revenues; IHS Technology (2014). A solar inverter is a component that converts the variable direct current output of a solar panel into a utility frequency alternating current that can be fed into the grid or used by an off-grid network.

2. Across the value chain, out of 200 solar start-ups funded with venture capital as of 2008, 56 firms went bankrupt and closed by the end of 2012, and 38 of them went bankrupt in 2012. Twenty-one other start-ups have engaged in M&A and restructuring; Wesoff, (2013).

3. I.e. 60% of new investment in solar power; see footnote 18 for a definition of asset finance; BNEF (2012).

4. Cross-border M&A in electricity generation from renewable energy amounted to USD 50.8 billion in 2012, with 281 operations over 2003-09 (USD 344 billion), concentrated in Brazil, China, India and Turkey; UNCTAD (2010).

5. I.e. trade as a percentage of total solar PV gross output in the solar PV sector; OECD (2010d).

6. Trade in Value Added (TiVA) describes a statistical approach used to estimate the source(s) of value (by country and industry) that is added in producing goods and services for export (and import). It recognises that growing global value chains mean that a country's exports increasingly rely on significant imports of intermediate goods; OECD (2013f).

7. i.e. determining where a product comes from. Such determinations are no longer simple, as raw materials and parts criss-cross the globe to be used as inputs in scattered manufacturing plants. Rules of origin are important in implementing such trade policy instruments as anti-dumping and countervailing duties, origin marking, and safeguard measures; WTO (2014c).

References

AFASE (2012), Solar Panels – A Global Value Chain – A Global Product, Alliance for Affordable Solar Energy, Brussels.

Barker, M. (6 February 2013), "What's next for solar trade disputes in 2013?", Solarbuzz, http://www.solarbuzz.com/resources/blog/2013/02/what's-next-for-solar-trade-disputes-in-2013.

BNEF (2013a), "New investment in clean energy fell 11% in 2012", Press Release of Bloomberg New Energy Finance.

BNEF (2013b), "2012 BNEF Clean energy & energy smart technologies league tables – The results", White Paper of Bloomberg New Energy Finance.

BNEF (2013c), BNEF Database, Bloomberg New Energy Finance.

BNEF (2013d), Value Chains, Bloomberg New Energy Finance client website, (accessed 27 August 2013).

BNEF (2012), "North-south clean energy investment flows: An $8bn step to a $100bn goal", Clean Energy White Paper of Bloomberg New Energy Finance.

Cao, J. and F. Groba (2013) "Chinese renewable energy technology exports: The role of policy, innovation and markets", Discussion Papers of German Institute for Economic Research, DIW, Berlin.Clean Energy Ministerial (2011), "Opportunities for economic value creation along the solar and wind value chain", Multilateral Working Group on Solar and Wind Energy Technologies input paper, Clean Energy Ministerial.

Clean Energy Ministerial (CEM) (2011), "Opportunities for Economic Value Creation along the Solar and Wind Value Chain", Input Paper, Multilateral Working Group on Solar and Wind Energy Technologies, CEM, 20 December 2011.

Curtin, J. (2013), "Comment: Solar PV trade wars threaten entire industry", RTCC (Responding to Climate Change).

De Gucht, K. (2012), "A trade defence system for today's global economy", speech of the European Commissioner for Trade, Brussels, 10 May 2012.

Deloitte/AEE (2011), Macroeconomic Impact of Wind Energy in Spain, Spanish Wind Association.

Ehrentraut O., et al (2013), The Impact of Anti-Dumping and / or Countervailing Measures on Imports of Solar Modules, Cells and Wafers from China on EU Employment and Value Added, Prognos AG, Basel, Berlin.

EPIA and Greenpeace (2006), Solar Generation: Solar Electricity for Over One Billion People and Two Million Jobs by 2020, European Photovoltaic Industry Association/ Greenpeace International.

EPIA and Greenpeace (2011), Solar Generation 6: Solar Photovoltaic Electricity Empowering the World, European Photovoltaic Industry Association/Greenpeace International.

EWEA (2012), Green Growth: The impact of wind energy on jobs and the economy, European Wind Energy Association, Brussels.

fDiIntelligence (2013), The fDi Report 2013: Global Greenfield Investment Trends, fDi Intelligence, London.

Frankfurt School-UNEP Centre, BNEF(2014), Global Trends in Renewable Energy Investment 2014, Frankfurt School of Finance & Management gGmbH.

Ghosh, A. and H. Gangania (2012) Governing Clean Energy Subsidies: What, Why, and How Legal?, ICTSD Global Platform on Climate Change, Trade and Sustainable Energy, International Centre for Trade and Sustainable Development, Geneva.

Goodrich, A. et al. (2011), "Solar PV Manufacturing Cost Analysis: U.S. Competitiveness in a Global Industry", National Renewable Energy Laboratory ("NREL").

Goossens, E. (2012), "Average U.S. Wind Farm Creates 1,079 Jobs, Report Finds", Bloomberg News.

GTM (2012), "PV Technology, Production and Cost: 2012-2016 Outlook", GTM Research.

GTM (2011), U.S. Solar Energy Trade Assessment 2011: Trade Flows and Domestic Content for Solar Energy-Related Goods and Services in the United States, Greentech Media.

Helpman, E. et al. (2004) "Export versus FDI with heterogeneous firms", the American Economic Review, vol. 94, No. 1.

Hufbauer, G.C. et al. (2013), Local Content Requirements: A Global Problem, Peterson Institute for International Economics, Washington, DC.

ICTSD (2011a), Fostering Low Carbon Growth: The Case for a Sustainable Energy Trade Agreement, International Centre for Trade and Sustainable Development, Geneva.

ICTSD (2011b), Feed-in Tariffs for Renewable Energy and WTO Subsidy Rules, Trade and Sustainable Energy Series, Issue paper 4, International Centre for Trade and Sustainable Development, Geneva.

IEA (2013a), *Tracking Clean Energy Progress, IEA Input to the Clean Energy Ministerial*, OECD/IEA, Paris. http://www.iea.org/publications/tcep_web.pdf

IEA (2013b), *PVPS Report: A Snapshot of Global PV 1992-2012*, Report IEA-PVPS T1-22:2013, OECD/IEA Publishing, Paris.

IEA/IRENA (2013a), *Solar Photovoltaics Technology Brief*, IEA-ETSAP and IRENA Technology Brief E11, the International Energy Agency and the International Renewable Energy Agency.

IEA/IRENA (2013b), "Global Renewable Energy Policies and Measures Database", the International Energy Agency and the International Renewable Energy Agency.

IRENA (2013a), *Renewable Energy and Jobs*, International Renewable Energy Agency, Abu Dhabi.

IRENA (2013b), *Renewable Energy Auctions in Developing Countries*, International Renewable Energy Agency, Abu Dhabi.

Kasteng, J. (2014), "Trade Remedies on Renewables: Intermediate and Long-Term Solutions", presentation at UNCTAD Clean energy and Trade Ad hoc Expert Group 2 on Trade Remedies in Green Sectors: The Case of Renewables, Geneva, 3-4 April 2014.

Kirkegaard, J. F. et al. (2010), "Towards a sunny future? Global integration in the solar PV industry", Working Paper Series, No. 10-6 (May 2010), WRI, Washington, DC.

Kirkegaard, J. F. et al. (2009), "It should be a breeze: Harnessing the potential of open trade and investment flows in the wind energy industry", Working Paper Series No. 09-14 (December 2009), WRI, Washington, DC.

Krüger, R. (2011), "Attracting foreign direct investment into renewable energy", presentation from UNCTAD at the Second International Energy Efficiency Forum Dushanbe, Tajikistan, 12 September 2011.

Lema, R. et al. (2012), "China's impact on the global wind power industry", Discussion Paper, 16/2012, DIE, Bonn.

Lema, R. et al. (2011), Competition and Cooperation between Europe and China in the Wind power Sector, IDS Working Paper 377, IDS, Brighton.

NRDC (2012), "Laying the foundation for a bright future: assessing progress under phase 1 of India's National Solar Mission", Interim Report, April 2012, Natural Resources Defence Council.

OECD (2013a), *OECD Code of Liberalisation of Capital Movements*, OECD Publishing, Paris, http://www.oecd.org/daf/inv/investment-policy/CapitalMovements_WebEnglish.pdf.

OECD (2013b), *Inventory of Estimated Budgetary Support and Tax Expenditures for Fossil Fuels 2013*, OECD Publishing, Paris, http://dx.doi.org/10.1787/9789264187610-en.

OECD (2013c), *Interconnected Economies: Benefiting from Global Value Chains*, OECD Publishing, Paris. http://dx.doi.org/10.1787/9789264189560-en.

OECD (2013d), *OECD Glossary of Statistical Terms*, http://stats.oecd.org/glossary/.

OECD (2013e), "OECD-WTO Database on Trade in Value-Added FAQs: Background Note", OECD, http://www.oecd.org/sti/ind/TIVA_FAQ_Final.pdf.

OECD (2013f), "The Climate Challenge: Achieving Zero Emissions", lecture by the OECD Secretary-General, Mr. Angel Gurría, London, 9 October 2013.

OECD (2013g), "OECD roundtable consultation on achieving a level playing field for international investment in solar and wind energy", Paris, 6 December 2013, Summary report.

OECD (2010a), "State owned enterprises and the principle of competitive neutrality", OECD Competition Committee, Policy Roundtables, OECD Publishing, Paris.

OECD (2010b), "Electricity: Renewables and Smart Grids", OECD Competition Committee, Policy Roundtables, OECD Publishing, Paris.

OECD (2010c), "Green growth and the crisis", OECD, Paris.

OECD (2010d), "Trade and the economic recovery: Why open markets matter", brief, OECD.

OECD (2010e), "Competition, state aids and subsidies", OECD Global Forum on Competition, Policy Roundtables, OECD Publishing, Paris.

OECD (2008), OECD Benchmark Definition of Foreign Direct Investment, Fourth Edition, OECD Publishing, Paris.

OECD, WTO and UNCTAD (2014), "Report on G20 trade and investment measures (mid-November 2013 to mid-May 2014)".

OECD, WTO and UNCTAD (2013a) "Implications of global value chains for trade, investment, development and jobs", prepared for the G-20 Leaders Summit at Saint Petersburg, September 2013.

OECD, WTO and UNCTAD (2013b), "Report on G20 trade and investment measures (mid-October 2012 to mid-May 2013)".

Platzer, M.D. (2012), "U.S. Solar Photovoltaic Manufacturing: Industry Trends, Global Competition, Federal Support", CRS Report for Congress, Congressional Research Service, Washington, DC.

Pew (2013), Who's Winning the Clean Energy Race?, 2012 Edition, The Pew Charitable Trusts.

REN21 (2014), *Renewable 2014 Global Status Report*, Renewable Energy Policy network for the 21st Century.

REN21 (2013), *Renewables 2013 Global Status Report*, Renewable Energy Policy network for the 21st Century.

Rutovitz, J. and A. Atherton (2009), "Energy sector jobs to 2030: a global analysis," prepared for Greenpeace International by the Institute for Sustainable Futures, University of Technology, Sydney.

Smith P., (2014), "Question of the week: local content requirements - help or hinder?", Wind Power Monthly, 20 June 2014.

The Solar Foundation (2014), *National Solar Jobs Census 2013: The Annual Review of the U.S. Solar Workforce*, The Solar Foundation, Washington, DC.

The Solar Foundation (2011), *National Solar Jobs Census 2011: A Review of the U.S. Solar Workforce*, The Solar Foundation, Washington, DC.

UNCTAD (2010), World Investment Report 2010: Investing in a Low-Carbon Economy, United Nations Publication, New York.

USITC (2011), "U.S. solar photovoltaic (PV) cell and module trade overview", USITC Executive Briefings on Trade, United States International Trade Commission, Washington, DC.

Wesoff, E. (2013), "Rest in Peace: The list of deceased solar companies", Greentech Media.

WTO (World Trade Organisation) (2014a), WTO Dispute Settlement Gateway.

WTO (2014b), "Joint statement regarding trade in environmental goods", Davos, 24 January 2014.

WTO (2014c), "Rules of origin", http://www.wto.org/english/tratop_e/roi_e/roi_e.htm.

WTO (2014d), Agreement on Subsidies and Countervailing Measures ("SCM Agreement").

WTO (1994), Agreement on Technical Barriers to Trade ("TBT Agreement").

Wu, M. and J. Salzman (2014), "The next generation of trade and environment conflicts: the rise of green industrial policy", Northwestern University Law Review, Vol. 108, No. 2, 410-474.

Chapter 3

Local-content requirements in the solar- and wind-energy global value chains

This chapter describes the increasing use of local-content requirements in green industrial policies. It discusses the arguments for and against such policies and describes the key findings of recent evidence-based analysis regarding their possible impacts on international investment in different segments of the global value chains. The analyses include; (i) a review of recent WTO disputes associated with the use of LCRs in solar and wind energy; (ii) an overview of recent investor-state disputes; (iii) results from a 2014 OECD Investor Survey assessing how leading international investors in wind and solar projects perceive the impacts of policies that differentiate between foreign and domestic investors; (iv) results from a consultation with key private and public stakeholders in solar and wind energy hosted by the OECD in December 2013; (v) results from a new econometric analysis conducted by the OECD to estimate the quantitative impact of LCRs on international investment flows to the solar- and wind-energy sectors; (vi) findings from other quantitative analyses of the impact of LCRs on international trade; and (vii) empirical evidence compiled from the literature on the effects of LCRs in several individual countries. Finally, the chapter discusses the policy implications of these research findings.

This chapter provides new empirical evidence to assess the arguments often raised in favour or against the use of LCRs in green industrial policies. As previously explained, LCRs are a type of localisation barrier to trade that requires solar- or wind-energy investors to source a certain share of inputs locally to be eligible for public support. Policy makers often claim that LCRs can help develop a domestic manufacturing base, create local jobs and promote technology transfer, in order to create local industrial clusters. Beyond the alleged or perceived economic benefits of LCRs, creating local content can help broaden the base of support for renewable energy incentive programmes. The political economy of support to renewable energy thus helps explain the prominence of LCRs in solar PV and wind energy. The benefits of LCRs, however, are heavily debated. Local-content requirements directly distort trade and may have some unintended effects on investment across value chains. Policies that favour some firms over others involve a cost, and can result in reduced competition and efficiency losses, thereby damaging the investment environment. Table 3.1 summarises the arguments on the possible impacts of LCRs. This chapter uses evidence-based analysis to assess whether LCRs are effective in achieving their intended policy goals. It also assesses the unintended impacts of LCRs, e.g. on international investment and competitiveness of solar and wind energy vis-à-vis fossil-fuel sources. Key findings are summarised at the end of the chapter.

This chapter builds on new empirical evidence, as well as existing studies and relevant OECD work (Cárdenas Rodríguez, M., et al. , 2014; Bahar et al., 2013, OECD, 2013a, c). The analysis includes:

- A review of recent WTO disputes associated with the use of LCRs in solar and wind energy;
- A brief overview of recent investor-state disputes;
- Results from a 2014 OECD Investor Survey assessing how leading international investors in wind and solar projects perceive the impacts of policies that differentiate between foreign and domestic investors, such as LCRs (see Annex 3.A2);
- Results from a stakeholder consultation hosted by the OECD in December 2013 with key private and public stakeholders in solar and wind energy (OECD, 2013g);
- Results from a new econometric analysis conducted by the OECD to estimate the quantitative impact of LCRs on international investment flows in solar and wind energy (Annex 3.A1);
- Findings from existing econometric analysis to model the impact of LCRs on international trade, including lessons learned from other sectors; and
- Empirical evidence from a review of the literature and recent case studies on the effects of LCRs in Brazil, Canada, China, India, South Africa and Ukraine.

Table 3.1 **Arguments on the possible impacts of local-content requirements in solar and wind energy**

Arguments and claims on the benefits of LCRs	Arguments and claims on the negative impacts of LCRs
Fostering nascent industries by protecting them from foreign competition until they achieve their latent competitive advantage	**Inefficient allocation of resources, trade diversion and distortion of competition. LCRs distort international trade** because they encourage the substitution of imports by domestic goods, even when their quality may be inferior and their price higher than those of foreign imports.
Providing medium-run economic spillovers by increasing the number of market players, which can lead to increased competition and innovation	**Reduced imports and competition** (i.e. market power) in the short-run between domestic manufacturers and foreign competitors. LCRs can **delay economies of scale and prevent cost reductions** for manufacturers by attracting high-cost firms and encouraging investment decisions based on public support, rather than on the cost-efficiency of specific locations
Learning spillovers (e.g. local training, technology transfer, knowledge and innovation), through learning-by-doing and capacity building	**Increased overall costs for downstream power producers in the short-run** as LCRs can force firms to purchase more expensive or less efficient solar panel or wind turbine equipment to benefit from public support
Economic diversification by creating business linkages locally	**Increased wholesale electricity prices in the short-term** to offset increased costs
Short-term benefits e.g. local job creation in manufacturing	**Limited capacity to create additional local green jobs in the short-term**
Improved public acceptance of policy support to renewable energy	**Higher technology risk in the short-run for downstream firms** forced to switch to less-known local technologies
Increased manufactured exports	**Increased cost of capital and restrained access to financing for project developers in the short-run** by lowering the bankability of projects forced to purchase less reliable domestic components
Increased local ownership and control of manufacturing capacity	**Reduced innovation and technology transfer from trade** for intermediate goods
Increased tax base for governments due to a larger manufacturing industry	**Higher revenue risk for downstream firms** in the short-run since the potential for governments to adopt LCRs makes the cost of components, and therefore profits, less predictable
Greater deployment of solar and wind energy in support of climate change mitigation	Reduced competitiveness and thus lower deployment of solar and wind energy vis-à-vis fossil fuels, detrimental to climate change mitigation
	Missed opportunities to support downstream services
	Increased policy uncertainty and investment risk

Sources: Kuntze and Moerenhout (2013); Bahar et al. (2013); OECD (2013a); Barker (2013); Stephensonn (2013); Hufbauer et al. (2013); Ghosh et al. (2013); ICTSD (2011a); Wu and Salzman (2014).

Stock taking of LCRs

Local-content requirements (LCRs) have increasingly been used by developed countries and emerging economies, both at national and sub-national levels, to favour domestic solar-PV and wind-energy manufacturers, especially since 2008 (OECD, WTO and UNCTAD, 2014; Bahar et al., 2013; Kuntze and Moerenhout, 2013; Stephenson, 2013). They often apply to all enterprises that operate domestically, whether national or foreign, or to domestic projects. Local-content requirements typically require components to be manufactured "locally" or "domestically", which is typically defined as taking place in a specific country or region, regardless of the firm's nationality. Since the early 2000s, LCRs have been planned or implemented at national or sub-national level in at least 21 countries, including in seven OECD countries, nine emerging economies and five developing countries. These measures have substantially increased over the past five years, according to recent OECD monitoring. Table 3.2 provides a list of LCR policies in solar and wind energy.

Governments have used LCRs in a number of ways in the solar and wind-energy sectors. They are typically imposed as a precondition for access to financial support schemes such as feed-in tariff (FiT) programmes or direct financial transfers, or as part of eligibility requirements in renewable-energy public tenders. When attached to FiTs, LCRs are set as eligibility criteria to qualify either for the full FiT, or for a possible variable part of the FiT and premiums provided on top of the base FiT. A few countries use different LCR ratios depending on the technology used in downstream installations (e.g. in India; Box 3.1). The LCR share is calculated based on renewable-energy components and equipment and varies considerably across countries. Table 3.2 shows that most LCRs in solar PV and wind energy have been used in association with FiTs (6 LCRs are attached to a FiT and 9 to a FiT premium) and public tenders (10 LCRs). From 2004 to 2009, China for instance included a 70% LCR in public tenders for wind power plants; bidders' responses on how they would comply with the LCR accounted for 20 to 35% of the bid evaluation scores (Kuntze and Moerenhout, 2013).

Some countries have also designed LCRs as eligibility criteria for direct financial transfers such as subsidised loans and loan guarantees from government agencies and national development banks. The 9th OECD, WTO and UNCTAD report on monitoring *G20 Trade and Investment Measures* mentions for instance the use of new LCR guidelines in order to access financing support from the Brazilian Development Bank (BNDES) (OECD, WTO and UNCTAD, 2013b; Bahar et al., 2013).

Box 3.1 Illustrative case study on LCRs in India

In India, the Jawaharlal Nehru National Solar Mission (2009) uses a competitive bidding process for new solar power tenders. Developers have to abide by a 60% LCR for projects using PV crystalline silicon (c-Si) cells and a 30% LCR for solar thermal and concentrated solar power, to qualify for the 25-year power purchase agreement (PPA) with a fixed FiT.

- **Under phase I (2010-13)** of the National Solar Mission, PV modules using thin-film technology were exempted from the 60% LCR, unlike projects using PV panels with c-Si technology. Since October 2012, only locally manufactured PV modules can qualify for the "Off-grid and Decentralized Solar Applications" support scheme (which provides a capital subsidy of 90% of the benchmark cost for solar-PV power projects below 100 kW). The Indian government reformed its LCR policy in Phase II to resolve reciprocal WTO disputes with the US.
- **During Phase II (2013-17)**, the auction for 750 megawatts (MW) of PV capacity included a mandatory LCR, to be eligible to receive Viability Gap Funding.[1] The auction is the first of its kind to offer grants (up to USD 300 million) that can be used to cover up to 30% of the costs of project development. Despite delays, the auction received a large amount of bids (58 developers bid for 122 solar projects).
- **Phase III (2017-22)** aims to achieve grid parity for solar power by 2022, with a 20 000 MW target of on-grid installed solar capacity, and a 2 000 MW target for off-grid solar installations.

1 The Indian Viability Gap Funding (VGF) Scheme uses credit enhancement to support public-private partnerships in infrastructure projects. It subsidises the capital cost whenever costs cannot be recovered via user fees, through a onetime construction capital grant; Ang and Marchal (2013).

Sources: Hufbauer et al., 2013; Stephenson, 2013; Kuntze and Moerenhout, 2013; Bahar et al., 2013; Freshfields, 2011; METI, 2013; WTO, 2014a; updated of September 2014.

Table 3.2 **Summary of local-content requirements (LCR) in wind and solar energy across OECD countries, emerging economies and developing countries**

Country (level)	Type of measure and sector(s)	Details of the LCR	Law / Regulation / Programme	Year of implementation of policy support	Year of implementation of LCR (Status)
Argentina (subnational)	Chubut province: FiT eligibility for wind	10% LCR in 1999, increased to 30% in 2001, 60% in 2003, 80% in 2005 and 100% in 2007	Provincial Law Nº 4389	1999 (FiT)	1999 (Ongoing)
Brazil (national)	Access to financing for wind	60% LCR under the PROINFA programme (Incentive Programme For Alternative Sources of Energy). The LCR was dropped in 2009 and replaced by a 60% LCR under BNDES' FINAME subsidised loan programme for wind turbine makers to participate in auctions in 2012. From 2013, manufacturers have to obtain locally at least three of the four main wind-farm elements (i.e., towers, blades, nacelles and hubs).	BNDES' PROFINA programme (2004-09) then BNDES FINAME loan programme	2009 (loans), 2012 (auction)	2009 (Ongoing)
	Access to financing for solar	BNDES also announced in Aug. 2014 forthcoming LCRs for solar projects in Brazil's upcoming first national solar-only auction on 31 Oct. 2014. Until 2017, projects receiving funding must use PV modules assembled in the nation using locally produced frames, with additional requirements and methodology to be introduced in 2017. Brazil also held its second regional solar energy auction in Nov. 2014 in Minas Gerais. LCR is announced for participation in the auction.	Besides the regular FINAME loan programme (5% interest rate), BNDES will offer solar projects loans through its Climate Fund (0.1% annual interest rate on amounts limited to 15% of the total investment)	2014 (Auction)	Forthcoming - 2017
Canada (subnational)	Ontario: FiT eligibility for solar and wind	50% to 60% LCR for FiT eligibility, in addition to a FiT bonus for products sourced locally, removed in Feb. 2014 to comply with WTO ruling after being temporarily lowered to 20-28%	Clean energy and Green Economy Act	2009 (FiT)	2009-2014
	Quebec: Tender eligibility for wind	3 tenders in 2003, 2006 and 2010, each with a 40-60% LCR based on equipment costs, expected 2014 tender with a 60% LCR including 30% in specific regions	Energy Strategy (2006-15); electricity supply contracts by Hydro-Quebec	2003 (Tender)	2003
China (national)	Tender eligibility for wind	70% LCR counting as 20-35% of total bid evaluation score	Renewable Energy Law (confirmed the LCR)	2003 (Tender)	2003 (Abandoned in 2009)
	Access to financing for wind	Grants under the "Special Fund for Wind Power Manufacturing" contingent upon LCR	Special Fund for Wind Power Equipment Manufacturing and Ride the Wind Program	2008 (Grants)	2008 (Abandoned in 2011)
Croatia (national)	FiT bonus eligibility for solar and wind	100% of FiT for 60% or more of LCR; or 93-99% of FiT for LCR under 60%	New Tariff System	2012 (FiT)	2012
France (national)	FiT bonus eligibility for solar	10% FiT bonus on EDF repurchasing price if 60% of added value is sourced in the EU and EEA. It was repealed by New Decree-law adopted in May 2014 and no longer has legal effect. Electricity producers which submitted a complete application for connection to the network before 10 March 2014 can still benefit from the increased tariff provided by the earlier decree-law.	Decree-law (repealed by New Decree-law adopted on May 8 2014)	2002, 2006 then 2010 (FiT)	2013 (repealed in 2014)

Greece (national)	FiT bonus eligibility for solar	10% bonus on top of the FiT if at least 70% of equipment costs derives from EU and EEA products	Law 4062/2012 (amending the 2010 "Photovoltaic and Renewable Energy Sources" Act)	2010 (FiT)	2012 (ongoing)
India (national)	Tender and FiT eligibility for solar	Phase I (2010-13): 60% LCR for silicon, 30% for CSP, 0% for thin-film; Phase II (2013-17): Batch I for implementation of 750 MW aggregate capacity of Grid Connected Solar Power Project launched stipulating 50% LCR for solar cells and modules to be used in the projects, eligibility to a Viability Gap Funding (VGF); Phase III (2017-22)	Jawaharlal Nehru National Solar Mission	2010 (FiT)	2010-22 (ongoing)
Indonesia (national)	FiT bonus eligibility for solar	20% FiT bonus for 40% or more of local content	Regulation of MEMR No. 17 of 2013	2009 (FiT)	2013 (ongoing)
Italy (national)	FiT bonus eligibility for solar	Bonus of up to 10% if more than 60% of EU and EEA content	Conto Energia IV and V	2008 (FiT)	2012-13 (FiT expired in July 2013)
Jordan (national)	FiT bonus eligibility for solar and wind	A FiT of USD 0.12 per kWh was introduced in December 2012 (the first in the MENA Region) by the local electricity regulatory commission, which includes a 15% LCR bonus for projects of full Jordanian origin	Renewable Energy and Energy Efficiency Law (REEL) passed in 2012	2012 (FiT)	2012 and ongoing
Malaysia (national)	FiT bonus eligibility for solar	Bonus determined by type of local components used	Malaysian Renewable Energy Bill	2010 (FiT)	2010 (ongoing)
Morocco (national)	Tender eligibility for solar	30% LCR for participation in solar auctions	Renewable Energy Law 13-09 and MASEN Law 57-09 adopted in 2010	2010 (Auction)	2010 (ongoing)
Russia (national)	Tender eligibility for solar and wind	First tender in 2013: 35-50% LCR for wind and solar-PV plants in 2013-2014, to be raised to 65-70% by 2016-20	Decree No. 449 on the Mechanism for the Promotion of Renewable Energy	2013 (Tender)	2013 (ongoing)
Saudi Arabia (national)	Tender eligibility for solar and wind	The King Abdullah City for Atomic and Renewable Energy (KA-Care) will open bids to develop 2 850 MW of renewable energy projects. The first bidding round will not include any LCR. In the second full-scale bidding round, bidders will need to satisfy LCRs of 50-60%	King Abdullah City for Atomic and Renewable Energy (KA-CARE) Act	2014 (Tender)	2014 (to be implemented in 2015)
South Africa (national)	Tender or FiT eligibility for wind and solar	First procurement round: 25% LCR (2012-2013); second round: 40% (2013-2014); to be raised to 60%	Green Economy Accord	2009 (FiT)	2012 (ongoing)
Spain (subnational)	Galicia, Navarra, Castile and Leon, and Valencia: FiT or tender eligibility for wind	Used for concession tenders in provinces; 70% LCR in Galicia and Navarra, to benefit from national FiT	From 1994 for Navarra. Galicia wind energy plan came into effect in 2004	1999 (FiT)	Abandoned in 2012-13 via retroactive changes: in 2012, moratorium on RE projects by suspending FiT pre-allocation procedures; and in 2013, retroactive cut of remuneration option for RE generation based on market price and FiT premium, for both existing and future projects

Table 3.2 **Summary of local-content requirements (LCR) in wind and solar energy across OECD countries, emerging economies and developing countries** *continued...*

Country (level)	Type of measure and sector(s)	Details of the LCR	Law / Regulation / Programme	Year of implementation of policy support	Year of Implementation of LCR (Status)
Turkey (national)	FiT bonus eligibility for solar and wind	Solar-PV projects and wind power projects that use locally produced equipment are eligible for additional feed-in tariff payments: up to 50% bonus on top of the FiT of 13.3 USD cent/kWh for solar and up to 50% bonus on top of FiT of USD 7.3 cent/kWh for wind	Turkish Renewable Energy Law	2004 then 2010 (FiT)	2010 (ongoing)
Ukraine (national)	FiT eligibility for solar and wind	15% before 2013, then 30% LCR in 2013, to be raised to 50% in 2014	Law of Ukraine on Electric Power Industry (Law 10183), NERC Resolution No 744 (August 2013)	2009 (FiT)	2013 (ongoing)
United States (subnational)	FiT bonus eligibility for solar and wind (Bonus of up to 20% on state FiT or resale price of electricity	California: bonus for 20% Californian state content	Self-Generation Incentive Program	2008 (FiT) 2006 for Washington	
		Massachusetts: bonus for 12% state content	Commonwealth Solar II Rebate Program	2012 (tax rebates)	2012 (Ongoing)
		New Jersey: bonus for 50% state content	Renewable Energy Manufacturer's Incentive Program	TBC	TBC
		Ohio: bonus for 20% Ohio content	Wind Production and Manufacturing Incentive Program	2007 (FiT)	Abandoned in 2007
		Washington: variable[1]	Renewable Energy Cost Recovery Incentive Program	2006 (FiT)	Ongoing
Uruguay (national)	Tender eligibility for wind	20% LCR for equity participation and 80% LCR for maintenance work	Decree on Renewable Energy	2010 (Tender)	2010 (ongoing)

[1] E.g. the state of Washington pays USD 0.36 per kilowatt-hour for electricity generated from solar power using panels manufactured in Washington state, but only 18 cents per kilowatt-hour if the panels were made in other US states or in China; Morris, 2012.

Sources: Compiled by authors; Bahar et al. (2013); Kuntze and Moerenhout (2013); Wu and Salzman (2014); Bellmann (2013); Morris (2012); IRENA (2013a); Creed and Kordvani (2013); IEA (2014b); updated as of 9 September 2014.

Assessment of the impacts of LCRs in the solar-PV and wind-energy sectors

The potential impact on international trade of WTO disputes

The use of LCRs in solar and wind energy has prompted several WTO disputes – five out of a total of 70 WTO disputes since September 2010 (as of July 2014; Table 3.3). Local-content requirements in solar and wind energy can indeed be challenged under WTO rules. Relevant international trade agreements include: the 1994 General Agreement on Tariffs and Trade (GATT; article III.4, .5 and .8.a); the Agreement on Trade-Related Investment Measures[1] (TRIMs, Article II and Annex); the Agreement on Subsidies and Countervailing Measures (SCM; Article III 3.1b); the Agreement on Government Procurement (GPA, Article XVI, Annex 4); and the North American Free Trade Agreement (NAFTA, chapter 11).

In particular:

- Local-content requirements can be considered as TRIMS, e.g. when they are attached to a FiT. *n* In solar and wind energy, LCRs have recently been challenged under the TRIMs Agreement, in complaints involving notably Canada, China, the EU, India and the US (Table 3.3). In May 2013, the WTO ruled against the LCR conditioning access to Ontario's FiT programme, under the GATT and TRIMs Agreement. This ruling has set a precedent for other similar cases (Box 3.2);

- LCRs may also qualify as prohibited subsidies[2] and be challenged under the SCM Agreement. The WTO, however, did not rule Ontario's LCR inconsistent with the SCM Agreement in 2013.

- When they are set as eligibility criteria for public tenders, LCRs are covered by the Agreement on Government Procurement. Though the WTO has some rules governing LCRs, there are important gaps in those rules. For instance, LCRs in public tenders are difficult to challenge. In addition, most of WTO rules (GATT, SCM and TRIMs) apply to goods and not services, which are covered by the General Agreement on Trade in Services (GATS).

Table 3.3 **WTO disputes relating to LCRs in solar and wind energy**

Complainant	Respondent	Subject of dispute	Date	Dispute Settlement Reference
US	India	LCR on certain types of solar cells and modules under Phase I of the Jawaharlal Nehru National Solar Mission. Supplementary consultations requested in Feb. 2014 on LCRs under Phase II	Feb. 2013	DS456
China	EU, Greece, Italy	Greece and Italy offering FiT bonus contingent upon LCR; EU's Renewable Energy Program	Nov. 2012	DS452
EU	Canada	Ontario's LCR under the FiT programme for solar and wind energy	Aug. 2011	DS426
US	China	Grants, funds, or awards conditional on LCRs under the "Special Fund for Wind Power Manufacturing"	Dec. 2010	DS419
Japan	Canada	Ontario's LCR under the FiT programme for solar and wind energy	Sept. 2010	DS412

Source: WTO (2014a), WTO Dispute Settlement Gateway – Chronological list of disputes cases, http://www.wto.org/english/tratop_e/dispu_e/dispu_status_e.htm
http://www.wto.org/english/tratop_e/dispu_e/dispu_status_e.htm, last accessed 2 July 2014.

Box 3.2 **Illustration of WTO disputes related the use of LCRs: the case of Ontario, Canada**

In Canada, the Province of Ontario enacted in 2009 an LCR policy requiring that project developers of wind and solar power systems wishing to benefit from Ontario's feed-in tariff (FiT) procure 50% (for wind projects of over 10 MW) or 60% (for solar projects of over 10 MW) of total project value in Ontario.

In September 2010, Japan issued a complaint to the WTO about the LCR, followed in August 2011 by a similar complaint from the EU. Japan and the EU argued that the LCR breached WTO rules by treating non-Ontarian manufacturers unequally relative to Ontarian manufacturers.

In December 2012, a WTO panel upheld the complaints that the LCR was inconsistent with the non-discrimination principles found in Article III of the GATT (national treatment) and Article II of the TRIMS Agreement. However, it ruled that the LCR did not qualify as a prohibited subsidy under Article 1.1 of the SCM Agreement. In May 2013, the WTO's Appellate Body dismissed Canada's appeal against the ruling. It was the first example of an LCR attached to a FiT that was ruled WTO-inconsistent.

The Ontario Government has since revised its FiT to comply with the WTO ruling, by removing LCRs from large procurements and lowering LCRs for small and micro-FiTs, and replacing them with other measures (e.g. procurement quotas), giving priority to municipalities and local communities.

Sources: Hufbauer et al. (2013); Stephenson (2013); Kuntze and Moerenhout (2013); Bahar et al., 2013; WTO (2014a, c).

The rise of investor-state disputes

In recent years, LCRs attached to renewable-energy feed-in tariffs have been involved in a series of investor-state disputes (IISD, 2012). Foreign investors involved in renewable energy projects in a country with an LCR may initiate international investment arbitration to seek damages, e.g. under provisions on performance requirements of bilateral investment agreements (BITs) or under international agreements such as the Energy Charter Treaty. In 2011 for instance, Mesa Power Group LLC, a US-based company, sent Canada a Notice of Intent to Submit a Claim to Arbitration under the North American Free Trade Agreement (NAFTA), in relation with the LCR attached to Ontario's FiT programme. Under articles of Chapter 11 of the NAFTA, the firm complained that the LCR provided more favourable treatment to a domestic company, and imposed prohibited "buy local" performance requirements (IISD, 2012).

The investor-state dispute settlement (ISDS) system addresses arbitration claims by foreign investors against governments under investment treaties (Gaukrodger and Gordon, 2012). It is often seen as a critical component of governments' efforts to enhance the credibility of the commitments they make in international investment treaties (Gaukrodger and Gordon, 2012). In some cases, ISDS has addressed challenged to public policies that modify the balance of contracts. In the absence of discrimination against foreign investors or expropriation, its application to the regulation of contracts varies between treaties and is often disputed. ISDS is an important tool in ensuring national treatment for foreign investors. However, while investment treaties are a key tool to protect foreign investors from discrimination, they have been increasingly criticised for allegedly providing preferential treatment to foreign investors over domestic investors with negative consequences for overall investment and competition (Australian Productivity Commission, 2010). Some major jurisdictions (the EU and the US) have taken steps to seek to limit such possible effects by stipulating that their treaties should not give foreign investors greater rights than domestic investors. Moreover, with some notable exceptions, investment treaties generally do not address market access; they are used only to protect established investments.

Results from the 2014 OECD Investor Survey

From April to June 2014, the OECD undertook a new Investor Survey on "Achieving a Level Playing Field for International Investment in Clean Energy".[3] The OECD Investor Survey was designed to provide a better understanding of what global investors perceive to be key policy impediments to international investment in solar PV and wind energy. According to results from the survey, LCRs stood out for international investors[4] as the main policy impediment to international investment across the solar and wind-energy value chain (Table 3.4). Annex 3.A2 provides detailed results of the OECD survey. OECD survey results are consistent with new results from the questionnaire of the SETI Alliance on "Global Sustainable Energy Trade Barrier", launched in co-operation with the OECD (SETI Alliance, 2014; Annex 3.A2).

Table 3.4 **Percentage of international investors from different segments of the solar and wind-energy value chains who identified LCRs as an impediment**

Solar-PV energy		Wind energy	
Upstream or midstream	Downstream	Upstream or midstream	Downstream
75%	**73%**	**70%**	**72%**

Note: Based on a sample of 42 international investor respondents. The results include responses from 8 international investors who operate across upstream, midstream and downstream segments.

Source: OECD 2014 Investor Survey; Annex 3.A2.

Unsurprisingly, a majority of international investors involved in downstream activities of the solar- and wind-energy sectors selected LCRs as an impediment (Table 3.4). The surveyed international investors demonstrated similar positions on LCRs whether they were from the solar-PV or wind-energy sectors. More unexpectedly, a majority of international investors involved in upstream or midstream activities also identified LCRs as an impediment (Table 3.4). This result suggests that LCRs can hinder investment across the value chains, not only in downstream segments. Respondents also reported that they encountered LCRs across 11 OECD, emerging and developing markets.

An important proportion of the surveyed international investors concurred that LCRs negatively impacted their company's activities in PV and wind-energy markets. The LCR policies notably resulted in:

- Higher investment risk and uncertainty (48% across segments; and 64% for international investors involved in upstream or midstream segments);
- Increased cost of intermediary inputs for downstream installations and services (41% in the downstream segment); and
- Discouraging international investors from investing in renewable-energy plants (46% in the downstream segment).
- Conversely, only 29% of international investors across the value chain responded that LCRs had encouraged them to invest in local manufacturing or to source their inputs locally.

The other policy restrictions surveyed included: administrative and technical barriers (42% of international investors identified them as an impediment); FDI regulatory restrictions (36%); other trade measures (e.g. import tariffs, trade remedies and burdensome custom procedures; 33%); and restricted access to financing (31%).

Key findings from the 2013 OECD Roundtable Consultation

On 6 December 2013, the OECD hosted a roundtable consultation on "Achieving a level playing field for international investment in solar and wind energy" with public and private stakeholders, and discussed the impacts of LCRs on international trade and investment in solar and wind energy (OECD, 2013g).

Participants had diverging views on the effects of LCRs on both international and domestic investment. A representative from a firm operating across the entire value chain referred to LCRs as a "double-edged sword", with a potential to drive investment up in component manufacturing and down in electric power. Private-sector participants noted that the impacts of LCRs depend on the segment of the value chain at which the LCR is applied and the size of the market:

- On the one hand, a few manufacturers operating in several countries argued that LCRs can help to attract international and domestic investment in large markets or regional hubs, especially when LCRs are backed by large tenders and ambitious development goals. They noted that LCRs can also support technology transfer. In addition, several participants highlighted the short-term effects of LCRs on local job creation. A few officials from OECD countries emphasised that many governments have implemented LCRs to expand the base of support for renewable-energy incentive schemes. Some manufacturers noted that the presence of renewable-energy manufacturing in a country can even help deploy renewable energy by increasing the political pressure on governments to maintain renewable-energy jobs and giving policy makers the necessary confidence to adopt ambitious renewable-energy targets.
- On the other hand, other manufacturers expressed the view that LCRs are not suited for small- and medium-sized countries, and are inefficient in the long-run. Representatives from downstream power generation companies argued that LCRs may reduce their cost-competitiveness and result in distorted competition and reduced technology transfer and innovation in solar and wind energy. Several public and private participants stressed

that the value added and employment resulting from LCRs are neither sustainable nor optimal. They emphasised that LCRs could hinder manufacturers' competitiveness in the long-run by making them reliant on discriminatory measures. Some project developers and government officials also reminded that the overarching goal of making solar and wind energy cost-competitive vis-à-vis fossil-fuel energy in the long-run can be better achieved through open and competitive markets.

Econometric analysis of the impact of LCRs on international investment flows

The OECD has undertaken several econometric analyses to assess the impact of public policies on international investment flows in renewable energy (see Cárdenas-Rodríguez et al. 2014; Haščič et al., 2015). The present analysis[5] extends this modelling approach to empirically analyse the effects of LCRs on international investment flows in solar PV and wind energy. Results indicate that while feed-in tariff (FiT) policies play an important role in attracting international investment flows in solar PV and wind energy, LCRs have a detrimental effect on global international investment flows in these sectors and hamper the effectiveness of FiT policies to which they are attached.[6] This effect is measured based on total international (or cross-border) investment flows in solar-PV and wind-energy generation between 2000 and 2011.[7] While LCRs are estimated to have a detrimental effect on international investment flows, their impact on domestic investment flows is not significant.

The econometric analysis builds on the approach found in the literature on gravity models. The OECD modelled international (or cross-border) investment flows in solar PV and wind energy as a function of bilateral economic relations between countries. Those relations are proxied by:

- Public interventions (including FiT and LCR policies);
- Geographic and socio-economic framework and market conditions; and
- Geographic distance and proximity in investment conditions.

To isolate the effect of LCRs from FiT policies, the analysis implements a fully interactive model that estimates the effect of policies separately and simultaneously. The model estimates the effects of LCR policies on both the decision of whether to invest, and on the volume of investment. The analysis is based on the full sample of worldwide international investment flows, domestic flows and total domestic and international investment flows in PV and wind-energy generation. The model distinguishes between the countries providing outward investment (i.e. the "source countries" of outflows) and the countries receiving inward investment (i.e. the "destination countries" of inflows). See Annex 3.A1 for a more detailed discussion of the model's empirical specification, estimation method and descriptive statistics.

Results suggest that FiT policies – when used without LCRs – have a significant and positive effect on both the investment decision and the volume of worldwide domestic and international investment inflows in destination countries. This evidence suggests that raising the ambition of support policies such as FiTs can effectively increase the likelihood and volume of investments in green electric power.

Conversely, the results show that **LCR policies have a significant and negative effect on international investment inflows, when LCRs are attached to FiT policies** in destination countries. This means that attaching LCRs to FiTs hampers the effectiveness of FiTs in attracting international investment inflows. This effect is even slightly stronger when estimating the effect of LCRs on both domestic and international investment flows. In addition, results show that LCRs have no effect on the likelihood or volume of domestic investment.

This modelling effort is the first attempt to measure empirically the effect of LCRs on international investment flows. The results discussed above have potentially important policy implications, calling into question the effectiveness of LCRs in attracting international investment in renewable-energy plants. Several conceptual points, however, need to be considered in interpreting these results:

- Most LCRs have been implemented relatively recently (mostly since 2008, i.e. in the later years of the estimation sample), thus, the effects might be stronger for estimations using more recent data (after 2011).
- The current model estimates the impact of policies using a static model, thus capturing only contemporaneous effects. However, there are likely dynamic reinforcements of these effects over time, indeed, barriers to international investments can slow down learning effects in new technologies, prevent economies of scale in investments and harm bilateral trade and economic relations.

Estimates of the impacts of LCRs on welfare, employment and trade from general equilibrium models

Review of relevant non-OECD studies: the case of Ontario, Canada

A number of modelling-based analyses find that LCRs have mixed or negative impacts on welfare, trade and employment in solar and wind energy (Böhringer et al., 2012; Rivers and Wigle, 2011). In particular, a recent study applies a computable general equilibrium (CGE) model[8] to estimate the labour market impacts of the feed-in tariff policy used by the Government of Ontario in Canada (Böhringer et al., 2012). The paper conducts a sensitivity analysis and scenario modelling to estimate the impacts of removing the LCR attached to the Ontario FiT programme on welfare, wages and employment. It finds that removing the LCR component of the programme would make the FiT less distortionary and reduce the welfare cost of the FiT programme by 30% compared with the existing policy, by removing a binding constraint on electricity generators. Removing the LCR also leads to reduced upward pressure on electricity prices, by allowing electricity generators to purchase non-domestic renewable-energy equipment. The paper concludes that in Ontario, while the LCR has a positive impact on employment in green sectors, the net impact of the LCR on wages and employment is likely to be negative because it results in higher electricity prices, which in turn results in higher production costs and a reduced real wage. The analysis estimates that the LCR (which has since been revoked) increased the annual welfare cost of the FiT policy by USD 340 million while creating 2 620 green jobs, corresponding to a social cost of the LCR of USD 130 000 per job. This study, however, uses a static model, which does not account for dynamic parameters of the broader policy or market environment.

Using an OECD Trade Model to draw lessons from other sectors

The OECD has conducted simulations utilising available information for a subset of LCRs to highlight the effects of LCRs on trade in those countries and sectors that have put LCRs in place since 2008 (OECD, 2015 forthcoming). The analysis considers only LCR policies that are binding, i.e. LCR policies which result in an actual fall in intermediate goods and services imports, and is not limited to LCRs applied in the solar and wind-energy sectors. The analysis applies the OECD new Trade Model, which is intended to improve analytical capacity to assess the impacts of trade policies both at, and behind–the–border. The OECD Trade Model uses a multi-regional computable general equilibrium (CGE) model that allows for up- and downstream linkages between industries as well as the tracking of trade flows along their regional dimensions.[9]

The results of the model show that LCRs reduce imports, not just in the implementing country's main trade partners, but across the world trading community. Domestic prices increase in those industries that are subjected to LCRs, which hurts the profitability of downstream firms that are deprived from potentially cheaper inputs. The LCR protects domestic production of the upstream or midstream sectors and causes production to shift towards these sectors and away from other domestic sectors. Simultaneously, the industry's production is shifted strongly to intermediates, which is the predominant segment where LCRs are applied, and away from other end-use and downstream activities. Thus, as intermediate imports fall, imports of final goods increase. This is an important effect, as policy makers often intend to use LCRs to develop a complete domestic industry, but end up with a domestic intermediate industry, at the expense of other sectors, downstream activities and final goods production in the targeted sector.

The model is also able to capture the extent to which these policies induce trade diversion. Results show that LCRs lead to an increase in exports from the affected sectors in those countries implementing LCRs, at the expense of other exporting countries. This is driven by the ability of firms to charge a lower price in the export market while benefiting from the price increasing effects of the LCR on the domestic market. However, the implementing countries as well as other exporting counties see an overall reduction in total export, which results in a global efficiency loss.

Evidence from selected countries

The impacts of LCRs on local manufacturing capability: the case of solar-PV energy in India

The findings of multiple studies have called into question the effectiveness of India's LCR policy in promoting local solar-PV manufacturing, in the first phase of the Jawaharlal Nehru National Solar Mission (Box 3.1; Shrimali and Sahoo, 2014; Sahoo and Shrimali, 2013; Johnson, 2013a, b; NRDC, 2012; Hufbauer et al., 2012; Ganesan et al., 2014). India's National Solar Mission has set two goals: (a) deploy solar-PV capacity to 200 GW by 2022 (using an auction-based feed-in tariff and public tenders); and (b) support domestic c-Si manufacturing (using a LCR on c-Si PV generation, which exempted thin-film technology in Phase I). The National Solar Mission has succeeded in scaling up PV installed capacity, from 18 MW in 2010 to 2.2 GW in 2013 (REN21, 2014; NRDC, 2012). Nevertheless, evidence shows that the LCR policy has had low effectiveness in deploying local c-Si manufacturing, and has distorted India's PV market. The LCR policy has created a bias towards thin-film imports (Hufbauer et al., 2013; Shrimali and Sahoo, 2014). More than 70% of India's solar-PV capacity now uses imported thin-film panels, which represent only 11% of global PV capacity. Studies estimate that the LCR policy has allowed domestic c-Si manufacturers to capture only 3-7% of India's solar market, much less than initially planned (Hufbauer et al., 2013). India's solar c-Si manufacturing base has become less competitive over time, with low capacity utilisation rate (10-15%), companies becoming bankrupt or in debt, or manufacturers surviving by diversifying (Shrimali and Sahoo, 2014; Johnson, 2013b). Low levels of capacity utilisation have also reduced innovation in the c-Si sector, by encouraging manufacturers to use low-cost assembling. Key impediments to the growth of local c-Si manufacturing included poor infrastructure, underdeveloped local supply chain, lack of raw materials and insufficient access to financing (NRDC, 2012).

The impacts of LCRs on input costs and electricity prices: the cases of solar PV energy in India and wind energy in Brazil, Ontario and Quebec

By requiring downstream investors and producers to source inputs locally, LCR experiences in Brazil, Canada and India (Table 3.2) show that LCRs can raise the cost of inputs if domestic goods and services are more expensive than imported ones. This can lead to increases in prices of electricity generated from solar and wind energy.

In the case of India's PV market for instance, the LCR policy has increased the cost of solar-PV systems for Indian developers, of up to USD 0.08 per watt. This "balance of system" penalty results from the lower efficiency of thin-film imports created by the LCR (Cimino, 2013). Compared with the global average spot price of c-Si modules, this increase corresponds to a 12% rise in the cost of solar modules effectively paid by producers, i.e. up to a 3% increase on the total costs of solar systems, as a result of the LCR policy (Hufbauer et al., 2013; Cimino, 2013). Expanding the LCR policy to cover all solar technologies, as the Indian government has been considering, would lead to a much steeper rise of total system costs, which could increase wholesale electricity prices (Hufbauer et al., 2013).

Similarly, studies estimate that the impact of LCRs on the costs of wind turbine systems in Ontario and Quebec will eventually increase the price of electricity generated from wind power. Assuming that the overnight cost[10] of onshore wind-energy plant can be used as a proxy for the price of wind turbines, and that the United States represents the baseline cost for onshore wind turbine power plants in North America, Hufbauer et al. (2013) estimate that Canadian wind turbines cost an additional USD 386 per kW of electric capacity to install than US wind turbines. The LCRs in Quebec and Ontario are responsible for around USD 200 million in additional capital costs. This represents about half of the difference between Canadian and US costs, which are 28% lower (Hufbauer et al., 2013; Cimino, 2013).

In Brazil's wind-energy sector, BNDES's large auction system and cheap loans have been successful in attracting FDI and deploying wind energy. BNDES's 60% LCR policy, however, has led to increased cost of wind turbine systems and significant delays in the installation of wind power plants (Kuntze and Moerenhout, 2013; Rennkamp and Fortes

Westin, 2013). In addition, the LCR is linked to the weight of the wind turbine components. This further increases the stringency of the LCR, and forces developers to source the majority of wind turbine towers locally. Despite Brazil's large market size and wind-energy potential, only a small share of wind-power developers decided to comply with the LCR in 2009 and 2010 tenders to benefit from BNDES financing (Kuntze and Moerenhout, 2013). The higher price of locally produced turbines stems from Brazil's high steel prices, which are around 70% more costly than imported steel and remain an important barrier to the profitability of wind-power projects in Brazil.

The impacts of LCRs on technology transfer and investment: the cases of wind energy in Brazil, China and Ukraine

Evidence from Brazil, China and Ukraine (Table 3.2) suggests that LCRs can hinder competitiveness, technology transfer and investment in wind energy, especially in the absence of sufficient technological capability, market size and financial support.

In Brazil, the LCR policy has incentivised the domestic production of low- and medium-technology content, but not of high-technology components of wind turbines. Wind-energy producers source locally heavy parts which are difficult to transport, and increasingly, parts of the nacelle, hubs and blades. Producers, however, continue to import expensive components with high-technology-content and high-quality jobs (Rennkamp and Fortes Westin, 2013).

The case of China's wind-energy sector demonstrates the benefits of removing LCRs once the industry matures, including on technology transfer and investment. From 2003 until 2009-11, several studies acknowledge that the LCR helped to develop China's nascent wind-turbine manufacturing industry (Xianqiang, 2013). Exceptionally favourable conditions included: the size of public tenders; substantial financial support from China; financing through the Clean Development Mechanism (CDM); China's large wind-energy potential; the market size of China's manufacturing sector; and growing domestic energy demand (Kuntze and Moerenhout, 2013). The LCR, however, deterred higher quality imports from more experienced foreign wind-energy producers (Xianqiang, 2013). Several foreign manufacturers left the Chinese market after 2003; it reduced opportunities for continued technical and operational improvement and limited quality improvement of Chinese wind turbines (Xianqiang, 2013; Kuntze and Moerenhout, 2013). Indeed, an LCR typically does not require the transfer of know-how or intellectual property rights; thus foreign manufacturers were not required to involve Chinese-owned firms in wind turbine design and assembly (Lewis et al., 2011). Conversely, since 2011, evidence suggests that the removal of LCRs in China's mature wind-energy sector (following negotiations with the US) has encouraged technology transfer, through imports of higher quality products (e.g. with Germany's Nordex contract to supply wind turbines for Chinese utility company Ningxia Electric Power). Following the removal of LCRs, there was also an increase in foreign investor's interest in investing in China (e.g. with Vestas and Gamesa in 2009).

In Ukraine, the increasingly stringent LCR (from 30% to 50%) component of the feed-in tariff has been extremely difficult to implement and has hindered international investment in wind energy (EWEA, 2013; ICTSD, 2013c). This is because Ukraine lacks industrial capability and local manufacturers able to produce relevant wind-turbine equipment. As of 2012, Fuhrländer Wind Technology was one of the only companies that could produce wind turbines locally. Weak industrial experience with MW-class turbines has made it difficult for companies operating in Ukraine to obtain domestic components in order to be certified from the National Electricity Regulatory Commission, and be eligible to the FiT. In addition, the lack of methodology and guidelines to calculate and comply with the LCR has further challenged the implementation of the LCR. Several studies estimate that the LCR has delayed or even prevented project developers from planning new renewable energy plants in Ukraine (OECD, 2012b).

Key findings and policy implications on the use of LCRs in solar PV and wind energy

Local-content requirements in solar and wind energy have increasingly been used as a component of "green industrial policies" to promote the development of domestic manufacturing in developed and emerging economies, particularly since 2009. Empirical analysis, however, suggests that LCRs have had mixed or negative effects on local job creation, value added and technology transfer in PV and wind-energy sectors. New empirical evidence also demonstrates that local-content requirements can hamper international trade and reduce competitive pressures on domestic manufacturers to lower their costs. By raising the cost of inputs for downstream businesses, LCRs can lead to increased overall costs, reduced price competitiveness less international investment, and increased wholesale electricity prices. Several conditions can further worsen the effects of LCRs on international trade and investment, and their ineffectiveness in achieving policy goals, including:

- Insufficient market size and local demand, or lack of regional co-operation (Johnson, 2013b);
- Insufficient domestic industrial experience and technical local expertise;
- Lack of access to financing;
- Lack of predictable, enforceable and flexible policy design of LCRs, to ensure they are time-limited, with planned evaluation phases to phase them out when the industry matures; and
- Design of LCRs that are not technology-neutral or that are too stringent and restrictive;

At the same time, creating value added and jobs locally is an important and ubiquitous goal, and facilitates public acceptance of policy support to clean energy. It is thus important for governments to assess the full effects of LCRs across the solar-PV and wind-energy sectors. In particular, policy makers could usefully recognise and assess the full potential for value creation and employment beyond the manufacturing segment of the value chain of wind- and solar-energy technologies, and especially in downstream activities (such as project development, installation, operation and maintenance). The relative importance of downstream activities may not be sufficiently recognised and included in political priorities.

By gathering insights into the possible impacts of LCRs, this chapter provides policy makers with evidence-based analysis with a view to informing the design of clean-energy support programmes and levelling the playing field for international investment in clean energy. Assessing the full costs and benefits of LCRs can help policy makers reconcile policy trade-offs and support domestic industries without restricting international trade and investment (Stepp and Atkinson, 2012). Local-content requirements are typically unnecessary in countries that have a competitive advantage in solar- and wind-energy manufacturing, and which can benefit from promoting open international trade in the sectors. In countries with a nascent or uncompetitive solar- or wind-turbine industry, policy makers can address local impediments that hinder the sector's competitiveness rather than impose more expensive and lower-quality panels and turbines on producers and consumers (EBRD, 2012). Improving the enabling conditions for investment in clean energy is likely to be more effective than imposing LCRs. Creating a stable and predictable policy environment for both domestic and international investment in clean-energy generation is critical, as emphasised by the OECD *Policy Guidance for Investment in Clean Energy Infrastructure* (OECD, 2015). Policy makers could usefully consider alternatives to LCRs to support their domestic solar-PV and wind sectors. In particular:

- Well-targeted R&D support in solar and wind energy can stimulate innovation in solar- and wind-energy manufacturing and encourage technology transfer from trade and FDI, without favouring domestic manufacturers. R&D support itself canl help build local manufacturing capability;
- Training programmes and promotion measures can improve the technological skills of midstream manufacturers, build local capability of downstream firms and encourage innovation across the value chains (Hufbauer et al., 2103; Johnson, 2013a, b; Kuntze and Moerenhout, 2013); and
- Demand-side instruments can help increase domestic demand without attaching a LCR, and can eventually support domestic manufacturing. Well-designed and predictable measures such as feed-in tariffs, auctions and tax incentives – or more cost-effective carbon pricing instruments – can also encourage wider deployment of solar and wind energy.

This report advises against the use of LCRs. Nonetheless, it acknowledges that there are views that LCR policies can be effective in achieving their policy objectives under certain conditions – e.g. sufficient market size and technical local expertise (Johnson, 2013a; Kuntze and Moerenhout, 2013). This report also recognises the political economy of LCRs, which can provide political benefits. In particular, they can: broaden the basis of support for renewable-energy incentive programmes; increase the political pressure on governments to maintain renewable-energy jobs; and give policy makers the confidence to adopt ambitious renewable-energy targets and support.

Notes

1. See footnote 7 p.43, Chapter 2.

2. Article 3.1(b) of the SCM Agreement prohibits subsidies contingent, whether solely or as one of several other conditions, upon the use of domestic over imported goods; Chapter 4.

3. The Survey was administered from April to June 2014 through an online questionnaire sent to leading global manufacturers, project developers, and financiers in the solar-PV and wind-energy sectors. Results are based on a sample of 62 respondents working for 59 companies involved across the upstream, midstream and downstream segments of solar-PV and wind-energy value chains.

4. I.e. investors who had already invested or participated in solar PV or wind-energy projects in foreign countries. In the survey sample, 68% (42) of respondents were "international investors".

5. Developed by Miguel Cárdenas-Rodríguez (OECD).

6. FiT policies include FiTs or a FiT premiums (or bonuses). The study does not consider LCRs attached to a loan or a tender.

7. This analysis uses the Bloomberg New Energy finance (BNEF) database to construct measures of international (i.e. cross-border) investment flows, including inflows and outflows. From this database, the OECD extracts 4 601 bilateral flows that: originate from 72 different countries and flow to 64 countries (representing most countries with a FiT policy); span the 2000-2011 time period; and cover investment in wind and solar PV power generation.

8. CGE models (also referred to as applied general equilibrium or AGE models) use actual economic data to estimate how an economy might react to changes in policy, technology or other external factors.

9. The model has two features that are unique for CGE models: (1) It integrates information from the OECD and WTO Trade in Value Added database (TiVA), which allows for a better assessment of the effects of policies on GVCs, using a more fully developed trade structure and a better reflection of the GVC trade patterns; and (2) it represents both price-based and quantity-based policy instruments. Most empirical trade models convert quantitative restrictions into tariff equivalents. However, price-based policy instruments lead to different adjustment processes than quantity-based measures, so it is important to distinguish between the two types of policy instruments. By allowing different policy-induced outcomes on domestic versus export markets, this approach allows firms the opportunity to engage in price discrimination (charging one price in the protected domestic market while charging another in the more competitive export market). Thus trade partners can lose twice: once through a decrease in their exports to the LCR-imposing country and again through a loss of market share in third-party markets from an increase in LCR-supported exports; OECD (2015 forthcoming).

10. I.e. the cost of preconstruction, engineering, procurement, construction and contingency costs; Hufbauer et al. (2013).

Annex 3.A1

Econometric analysis of the impact of local-content requirements on international investment flows

The OECD has undertaken several econometric studies to analyse the impact of policies on international investment flows in renewable energy sectors (Cárdenas-Rodríguez et al., 2014; Haščič et al., 2015). The present analysis, which is authored by Miguel Cárdenas Rodríguez from the OECD Environment Directorate,[1] extends this modelling approach to empirically analyse the effect of local-content requirements (LCRs) on international investment flows in solar PV and wind energy. Results indicate that while feed-in tariff (FiT) policies play an important role in attracting both domestic and international investment inflows, LCRs hamper the effectiveness of FiT policies[2] in attracting international investment inflows in solar PV and wind energy, when LCRs are attached to FiT policies. This effect is even stronger when considering the impact of LCRs on worldwide flows (both international and domestic investment).

Econometric techniques allow for the estimation of partial correlations between the occurrence of investment and individual factors relating to a range of public interventions (finance, policies, and measures) as well as broad market and country conditions. For example, these techniques allow the effect of each policy in a policy mix to be isolated (e.g. LCRs attached or not to a FiT). Moreover, they provide a basis for investigations of spillover effects and testing for causality. Most importantly, they allow for generalised policy conclusions to be drawn. However, these techniques also face their own set of difficulties, such as the level of detail that can be accounted for; required variation in data; potential identification problems; sample selection issues as well as data availability and quality more broadly. This means that if decision-makers seek empirical guidance to estimate the mobilisation impact of broadly-defined and widely applicable public interventions (e.g. LCRs and feed-in tariffs in generic terms), then econometric studies have the potential to provide value added. However, for guidance and estimation of mobilisation resulting from public interventions that are relatively less frequent or are placed in a highly specific context (e.g. renewable-energy tenders), conducting a case study remains the only option.

The study uses the Bloomberg New Energy Finance (BNEF) database (BNEF, 2013c) to construct measures of international (i.e. cross-border) investment flows including inflows and outflows). From this database, the OECD extracts financial deals that: originate from 72 different OECD, emerging and developing countries; flow to 64 OECD, emerging and developing countries; span the 2000-2011 time period; and cover investment in wind and solar-PV power generation.[3] The current lack of corresponding data on the domestic policy framework conditions prevents extending the time period to more recent years.[4]

For the purposes of this report, the model isolates the effect of LCRs associated with FiT policies on international investment in solar-PV and wind-energy generation. Pending data availability, future analyses could apply a similar method to an expanded set of both sectors and types of public interventions.

Empirical specification

The construction of the model and its empirical specification was inspired by the literature on gravity models. Gravity models were originally developed to study international trade and subsequently applied to study the impacts of trade on the environment, to study international technology transfer as well as international investment. Building on this approach and the recent OECD report "Public Interventions and Private Climate Finance Flows: Empirical Evidence from Renewable Energy Financing", this analysis models renewable energy investment as a function of bilateral (between pairs of countries) economic relations proxied by geographic distance, proximity in investment conditions, policy conditions, market conditions, etc. (Haščič et al., 2015). Formally,

$$Investment_{ijkt} = \beta_0 + \beta'_1 \boldsymbol{POLICY}_{ijkt} + \beta'_2 \boldsymbol{MARKET}_{ijkt} + \beta'_3 \boldsymbol{CONTROL}_{ijkt} + \alpha_i + \gamma_j + \delta_k + \theta_t + \epsilon_{ijkt} \quad [1]$$

Where $i = (1, \ldots, I)$ and $j = (1, \ldots, J)$ index respectively the source and destination country,[5] $k = \{1, 2\}$ indexing the sectors studied (wind, and solar PV) and $t = (2000, \ldots, 2011)$ indexes the year of the financial flow. As explanatory variables, the model includes vectors (in bold) that describe the policy and the market conditions alongside a vector of control variables.

$\boldsymbol{POLICY}_{ijkt}$ is a vector of variables that capture public interventions that are hypothesised to have an effect on renewable energy investment. Policy interventions supporting renewable energy are represented by two variables: feed-in tariffs (in USD per kWh using 2011 prices) for both source and destination country (FIT_{ikt} and FIT_{jkt} respectively) and renewable energy quotas (in percentage points) in source and destination country (REQ_{it} and REQ_{jt}). Both policy measures are taken from the OECD-EPAU (2013) Renewable Energy Policy Database (for a detailed discussion see Cárdenas Rodríguez et al. 2014). Next, the study constructs dichotomous (i.e. discrete rather than continuous) variables indicating the existence of LCRs in both source and destination countries (LCR_{ikt} and LCR_{jkt} respectively). The LCR variable is constructed based on an estimation sample of LCRs attached to feed-in tariff (FiT) programmes, FiT bonuses or premium and public tenders implemented in OECD countries, emerging economies and developing countries. The measures were implemented in the period from 2000 until 2013 using official sources and a review of the literature.[6] The countries where the measures were implemented include Australia, Belgium, Brazil, Canada, China, France, Germany, India, Italy, Japan, Jordan, Morocco, Portugal, South Africa, Spain, Sri Lanka, Uganda, Ukraine, the United Kingdom and the United States. Only dichotomous variables were used instead of actual percentages.[7]

$\boldsymbol{MARKET}_{ijkt}$ is a vector of geographic and socio-economic (framework) conditions that might influence financial flows between a pair of source and destination countries. The model uses the CEPII dataset[8] and include variables such as $Distance_{ij}$ measuring the geographic distance in kilometres between the most important cities of both countries, weighted by population. The model also includes dummy variables for $Common\_language\_d_{ij}$ (equal to one if both countries share the same official language and zero otherwise) and $Common\_legal\_system_{ij}$. Differences in official languages and legal systems might translate into higher costs for investors. Finally, a dummy variable is constructed that takes on the value of 1 when both the source and destination countries are jointly members of the World Trade Organization.[9]

The model also includes GDP per capita data from the World Bank database, for both source and destination country (GDP_{it} and GDP_{jt}). From the OECD/IEA World Energy Balances database, it also includes $Electricity\_consumption\_growth_{it\ (or\ jt)}$ of the source or destination country to reflect changes in market opportunities.

The volume of financial flows between countries is also a function of certain financial conditions in the respective countries (e.g. presence of a structured financial services industry, information asymmetries, exchange rate differences, trade protectionism or membership in regional trade agreements and the resulting patterns of trade and investment). To capture the influence of such bilateral investment conditions, a variable accounting for levels of Foreign Direct Investment (FDI_{ijt}) could be a suitable proxy. This would in particular help disentangle the financial motivation from the "environmental" motivation to invest in a given country. OECD and IMF were considered as potential data sources. However, FDI data is measured and aggregated in a particular manner (as a change in value of asset holdings, or "net" flows, with disinvestments accounted for as negative flows). This is not suitable for the purpose of this study which requires a measure of the volumes of new additional investment, or "gross" flows (only positive flows) in each direction, to be consistent with the construction of our dependent variable.

$\boldsymbol{CONTROL}_{ijkt}$ is a vector of variables that are intended to capture the idiosyncratic characteristics of the BNEF database. $BNEF\_Finance_{it}$ is constructed as the sum of all flows going out of country i while $BNEF\_Finance_{jt}$ are all the inward flows received by country j, including domestic flows. In constructing these two variables the flows are aggregated across all countries and all sectors. To avoid endogeneity problems, the analysis subtracts the amount of the dependent variable from these two control variables. These variables play an important role in the regression; they allow to (i) account for any possible biases due to idiosyncratic differences in BNEF database with respect to its coverage of countries, years, sectors and financial instruments; and to some extent (ii) control for the size of the clean-energy component of the financial market in both the source and the destination countries.

Finally, the model includes dummies for source country α_i, destination country γ_j, sector δ_k and year θ_t to capture any heterogeneity that is invariant in the respective dimension. The remaining variation of the dependent variable is captured by the error term ϵ_{ijkt}.

Estimation method

Models with dependent variables censured at zero are typically implement using a **Tobit** estimation procedure. A Tobit model is a mixed model associating: (1) a Probit that models the binary decision of investment, i.e. estimates the probability of observing a strictly positive flow of investment; and (2) a classic linear model (ordinary least squares regression on the uncensored observations) that analyses the amount invested once a positive investment decision has been taken. While Tobit analyses both (1) the investment decision as well as (2) the volume invested, it does so by assuming that that decision to invest and the volume invested are determined by the same process. Therefore, the analysis relaxes this assumption and implements a type-II Tobit (also called the "Heckman selection model") that allows to model two separate but correlated processes. To test the statistical significance of estimated coefficients, the analysis uses cluster-robust standard errors to account for possible heterogeneity across country-pair clusters.[10]

This estimation strategy gives a high importance to the procedure in which zeros for the dependant variable were imputed.[11] Indeed, a zero in our dataset, reflects the assumption that no investment for that combination of dimensions (*i,j,k,t*) existed. However, the absence of value may reflect the incomplete coverage of our dataset. As a consequence, the analysis opts to impute zeros on the basis of dimensions *i,j,t* (not *k*). This means that "if a sector is covered for a given country-year combination, then all sectors are covered". Consequently, the study imputes a zero investment volume for the remaining sectors (of the country-year combination). In our judgement, this is the most conservative approach. The idea is to compare the policy framework in countries in which investment occurs for a given year and sector against countries with no investment. Running the regression analysis without making any assumptions on zero investment would not let us test the "crowding-in/out" effect of policies.[12]

In order to isolate the effect of LCRs from FiTs, the study implements a fully interacted model (LCRs × FiTs) which will allow to estimate the effect of policies separately and simultaneously. The expectation is that such instruments have a negative effect on investment. Furthermore, they might hamper the ability of related public policies, such as feed-in-tariffs, to attract investment in renewable energy.

We obtain a maximum sample size of 4601 observations, including 769 country pairs (74 different source and destination countries), and covering the 2 renewable energy sectors and the 2000-2011 period. Table 3.A1.1 provides the descriptive statistics.

A **more detailed discussion on the econometric methodology is developed in Haščič et al. (2015)**. That report provides detailed insights on:

- The construction of the model specification;
- The choice of the estimation method;
- The relevant literature review and comparative assessment of the methods that can be used to estimate gravity models applied to the set-up;

- The choice of variables, inclusion of dummies and the choice of the exclusion variable appropriate to the estimation of Heckman selection models; and
- Robustness checks. A set of robustness checks was also performed in the line of the Haščič et al. (2015) paper.

Empirical results

First, the study estimates a model specification on a full sample of cross-border investment; second, it investigates the effect of LCRs on domestic flows and finally, it explores if results hold for flows worldwide.

The results for the worldwide sample suggest that *FiT policies* in destination countries play an important role (positive and significant coefficient) for both the investment decision and the volume of investment. This evidence suggests that if countries seek to encourage and effectively increase the likelihood and volume of investments, raising the ambition of policies in destination countries will be vital. On the other hand there is no evidence of the effect of FiTs in source countries, as they are not significant on the decision to invest neither on the volume of investment.

The core result of our model is presented when looking at the effect of *LCR policies* in source and destination countries, together with the interaction terms. First, *LCR policies* in the source country are correlated with lower volumes of outflows; however, when LCRs are attached to a FiT we observe a higher likelihood and volume of investment outflows. Second, *LCR policies* in destination countries do not show a significant effect on volumes nor likelihood of investment; nevertheless, when the LCR are combined with FiTs in destination countries, we see a significant and negative effect of LCRs for cross-border investment, and this result holds in the worldwide sample.[13] This means that attaching LCRs to FiTs hampers the effectiveness of FiTs to attract cross-border investment, and this effect is slightly stronger when we evaluate our model in the worldwide sample.

The evidence regarding *REQ policies* in source or destination countries is somewhat mixed. Concerning the impact on investment volumes, it seems that REQ policies in source countries are negatively correlated with the volume of investment outflows. On the other hand, *REQ policies* in destinations do not have an impact on the volume nor decision to invest; we see however that this effect is in fact positive when estimated in the domestic sample.

Results for the other explanatory variables vary to a certain extent. In most cases we find evidence that investors tend to invest in nearby destinations rather than in countries geographically far away, and in countries with which they share a common legal system (effect on the decision of investment), but not necessarily a common official language (which has an effect on volumes but not investment decision). We also find no significant effect of regional trade agreements or opportunities arising from the growth of electricity markets. On the other hand, a common legal system is an important determinant for the decision of investment. In terms of differences across sectors, we find a strong effect indicating that overall solar-PV investment is less likely to attract investment, and also is correlated with lower volumes of investment compared to wind power.

Finally, both BNEF control variables are often positive and significant suggesting that we successfully control for database coverage and hence partly mitigate idiosyncratic biases. We are for example more likely to observe positive and larger investment flows if the underlying database has a good coverage of this particular country-pair and year combination.

Conclusion

This modelling effort is the first attempt to empirically measure the effect of LCRs on international investment flows. The results discussed above have potentially significant policy implications. First, supporting policies, such as feed-in-tariffs, are attracting inflows of international investment. Second, the analysis shows LCRs have a strong negative effect when they are attached to policies supporting renewables (e.g. FiTs). This means that LCRs are not only preventing the inflows for key sectors such as solar-PV and wind-power infrastructure, but are also hampering the effectiveness of other policies that aim at promoting them. Overall, this effect can be seen for international investment, and it is even stronger for worldwide flows of investment, i.e. worldwide investment (both domestic and international) in solar PV and wind energy has decreased with the introduction of LCRs.

Several conceptual points need to be considered in the interpretation of these results. First, LCRs have been implemented relatively recently (mostly since 2008, i.e. only in the later years of the estimation sample), hence effects might be stronger for estimations using more recent data. Second, the current model estimates the impact of policies in a static model capturing only contemporaneous effects. However, there are likely dynamic reinforcements of these effects over time. Indeed, barriers to international investments can slow down learning effects in new technologies, prevent economies of scale in investments and harm bilateral economic and trade relations.

Table 3.A1.1 **Descriptive statistics**

Variable	Observations	Mean	Std. Dev.	Min	Max
Investment (mln USD)	4601	116.9533	1011.042	0	39027.7
Investment (ln)	1394	3.967902	1.971926	-3.0293	10.57203
Domestic Flow (dummy)	4601	0.173441	0.378669	0	1
FIT source	4601	0.122329	0.210535	0	1.598654
LCR source	4601	0.169963	0.375641	0	1
FIT destination	4601	0.114513	0.207167	0	1.598654
LCR Source	4601	0.142578	0.34968	0	1
REQ source	4601	0.021801	0.036906	0	0.182
REQ destination	4601	0.019898	0.036223	0	0.182
GDP per capita source (USD)	4601	215104.8	321215.8	208.5838	1113572
GDP per capita source (ln)	4601	11.29026	1.515138	5.340341	13.92308
GDP per capita destination (USD)	4601	169364.2	281649.4	554.4414	1113572
GDP per capita destination (ln)	4601	10.95371	1.599728	6.317961	13.92308
Weighted distance	4601	4499.346	4804.067	0	19516.56
Weighted distance (ln)	4601	6.657329	3.23089	0	9.87907
Common Official Language (dummy)	4601	0.346229	0.475819	0	1
Regional Trade Agreement (dummy)	4601	0.61052	0.487686	0	1
Common legal system (dummy)	4601	0.437731	0.496161	0	1
Electricity consumption growth in Source	4601	0.013027	0.041808	-0.123743	0.336512
Electricity consumption growth in Destination	4601	0.018463	0.043779	-0.124137	0.253386
BNEF control, source country (mln USD)	4601	7893.142	15800.33	0.120104	94069.28
BNEF control, source country (ln)	4601	7.387305	2.152699	-2.1194	11.45179
BNEF control, destination country (mln USD)	4601	5869.915	12657.52	0.1153	70038.15
BNEF control, destination country (ln)	4601	6.820832	2.268855	-2.160222	11.1568
Country-pair in WTO (dummy)	4601	0.983699	0.126644	0	1

Note: ln: log; mln: million.

Table 3.A1.2 **Estimation results**

	Full Sample (M1)	Cross-Border (M2)	Domestic (M3)
Volume of investment (ln) (linear equation)			
Domestic flow	-0.6274	–	–
	[0.6946]	–	–
FIT source	0.4089	-0.0056	–
	[0.2660]	[0.2828]	–
LCR source	-0.5219***	-0.6814***	–
	[0.1947]	[0.2123]	–
FIT Source × LCR Source	8.5382***	7.5976***	–
	[1.6880]	[2.0079]	–
FIT destination	1.3590***	0.7975	2.2076**
	[0.4683]	[0.4964]	[1.0209]
LCR destination	0.3742	0.185	-0.0791
	[0.3280]	[0.3587]	[0.4359]
FIT Destination × LCR Destination	-7.0939	-3.7677	-2.4283
	[4.3374]	[4.5436]	[5.4994]
REQ source	-3.9220**	-2.4134	–
	[1.8191]	[1.9732]	–
REQ destination	2.5374	3.0493	2.5839
	[3.9689]	[5.1140]	[5.8392]
GDP-per-capita source	0.3213***	0.1208	–
	[0.0664]	[0.0764]	–
GDP-per-capita destination	0.9324	0.8464	2.8541*
	[0.8433]	[0.7714]	[1.5823]
Weighted geographic distance	-0.4048***	-0.3699***	–
	[0.0991]	[0.1094]	–
Common official language	-0.2414	-0.3536**	–
	[0.1719]	[0.1667]	–
Regional Trade Agreement	-0.0506	-0.3106	–
	[0.2376]	[0.2294]	–
Common legal system	0.0778	0.0612	–
	[0.1288]	[0.1345]	–
Growth in electricity cons. source	-0.5374	-4.7155**	-1.7929
	[2.1699]	[2.0107]	[3.9597]
Growth in elect. cons. destination	0.3169	-0.5987	–
	[1.8966]	[1.7696]	–
BNEF control source	0.2329***	0.2432***	-0.0784
	[0.0495]	[0.0571]	[0.1288]
BNEF control destination	0.1263*	0.1770**	0.2933**
	[0.0697]	[0.0692]	[0.1434]
Solar PV sector (dummy)	-1.9325***	-1.2308***	-2.7006***
	[0.2362]	[0.2529]	[0.4977]
Year dummies	yes	yes	yes
Sector dummies	yes	yes	yes
Source country dummies	no	no	–
Destination country dummies	yes	yes	yes

Note: ln: log; mln: million.

Table 3.A1.2 **Estimation results...** *continued*

	Full Sample (M1)	Cross-Border (M2)	Domestic (M3)
Domestic flow	-0.0233	–	–
	[0.3372]	–	–
FIT source	-0.0883	-0.2349	–
	[0.1733]	[0.1949]	–
LCR source	-0.1293	-0.1109	–
	[0.0997]	[0.1088]	–
FIT Source × LCR Source	4.8714***	5.0526***	–
	[1.0986]	[1.1637]	–
FIT destination	0.5140**	0.4350*	0.9441***
	[0.2231]	[0.2368]	[0.3402]
LCR destination	0.0837	0.0845	0.0482
	[0.1720]	[0.2175]	[0.2624]
FIT Destination × LCR Destination	-4.4057**	-4.3165*	3.5385
	[2.0219]	[2.3092]	[3.9362]
REQ source	-1.8546**	-2.3647***	–
	[0.8499]	[0.8676]	–
REQ destination	-0.7713	-0.8962	5.9985***
	[1.8929]	[2.2204]	[2.0582]
GDP-per-capita source	0.2052***	0.2060***	–
	[0.0318]	[0.0354]	–
GDP-per-capita destination	0.6151	0.63	0.3659***
	[0.4046]	[0.5363]	[0.0825]
Weighted geographic distance	-0.1570***	-0.1717***	–
	[0.0462]	[0.0479]	–
Common official language	-0.0899	-0.1149	–
	[0.0848]	[0.0849]	–
Regional Trade Agreement	0.1027	0.0411	–
	[0.1098]	[0.1123]	–
Common legal system	0.1517**	0.1525**	–
	[0.0669]	[0.0659]	–
Growth in electricity cons. source	-1.6814*	-2.1725**	–
	[0.8588]	[0.8929]	–
Growth in elect. cons. destination	1.2466	0.7798	-0.7537
	[0.9235]	[1.0196]	[1.8205]
BNEF control source	0.0733***	0.0431	0.1575**
	[0.0252]	[0.0278]	[0.0695]
BNEF control destination	0.0337	0.0552*	0.1195**
	[0.0246]	[0.0305]	[0.0527]
Country-pair in WTO (dummy)	0.6408	0.8318*	–
	[0.4970]	[0.5001]	–
Destination countries in WTO (dummy)	–	–	1.3989***
	–	–	[0.2289]
Solar PV sector (dummy)	-0.6506***	-0.5740***	-0.9166***
	[0.0865]	[0.0967]	[0.2046]
Year dummies	yes	yes	yes

Table 3.A1.2 **Estimation results** *continued...*

	Full Sample (M1)	Cross-Border (M2)	Domestic (M3)
Sector dummies	Yes	yes	yes
Source country dummies	No	no	–
Destination country dummies	Yes	yes	no
Observations (of which uncensored)	4601 (1394)	3803 (1067)	798 (327)
Test of independence of equations (Null hypothesis that Rho=0)	rejected at 1%	not rejected	rejected at 1%

Notes

1. miguel.cardenasrodriguez@oecd.org.

2. I.e. That is to a feed-in tariff programme or a FiT premium or bonus. The analysis does not consider LCRs attached to a loan or a tender.

3. Thus excluding financial flows from solar-PV and wind- turbine manufacturing, which represent a small share of total investment flows in solar PV and wind energy; globally, equipment manufacturing represented only 6% of new investment in renewable energy in 2013; FS-UNEP and BNEF, 2014.

4. Lack of data also prevents from including other climate-related sectors, although they are not relevant to this report.

5. Flows to multilateral organisations are excluded from all estimations.

6. Compiled by authors; Bahar et al. (2013); BNEF (2013c); Kuntze and Moerenhout (2013); Wu and Salzman (2014); Bellmann (2013); Morris (2012); IRENA (2013a, 2013b); Creed and Kordvani (2013); IEA (2014b); updated as of July 2014.

7. In order to maximise the estimation sample with respect to the number of countries included in the analysis, the analysis opted to impute the LCR variable for additional countries not included in the dataset presented in Table 3.A1.1. This imputation relies on the assumption than such countries do not have LCRs in place. The analysis tests whether this imputation affects results by including a dummy variable indicating the imputed observations and running the analysis described below. The dummy for data imputation is never significant; hence, the assumption that these countries do not have LCRs does not affect results.

8. Harmonised data for gravity equations from the Centre of Prospective Studies and International Information (CEPII in French). http://www.cepii.fr/CEPII/fr/bdd_modele/bdd.asp

9. For a more detailed definition of these variables, please refer to Haščič et al. (2015).

10. Identification of the Heckman procedure is achieved with the proper selection of the exclusion variable. The study identified WTO membership for the country pair as a suitable exclusion variable: it is correlated with the decision of investment and uncorrelated with the volume. Haščič et al. (2015) explores this subject in depth.

11. To some extent, Heckman selection models are more suitable to account for excess of zeros than Tobit methods.

12. For a more detailed discussion of the imputation strategy refer to Haščič et al. (2015).

13. In fact, the results in the Model (M3), show that LCRs have no effect on domestic investment, neither likelihood nor investment volume.

Annex 3.A2

The view from private investors: Results from the 2014 OECD investor survey on "Achieving a Level Playing Field for International Investment in Clean Energy"

This annex summarises results from the 2014 OECD Investor Survey on "Achieving a Level Playing Field for International Investment in Clean Energy", based on a sample of 62 respondents from 59 leading companies involved across the solar-PV and wind-energy value chains. This OECD Investor Survey was undertaken to gain a better understanding of what private investors perceive to be key policy impediments to international investment in solar PV and wind energy. Results of the OECD Investor Survey offer insights to support the design of more effective policies for attracting international investment in solar PV and wind energy.

Methodology

The OECD Investor Survey was administered between April and June 2014, through an online questionnaire sent to leading global manufacturers, project developers, and financiers in the solar-PV and wind-energy sectors. The questionnaire was developed to assess the main concerns of international investors in solar PV and wind energy. It included inputs from key private-sector representatives in solar PV and wind energy, who participated in the OECD roundtable consultation on "Achieving a Level Playing Field for International Investment in Solar and Wind Energy" on 6 December 2013 (OECD, 2013g). Responses to the questionnaire were collected using the OECD Online survey tool.[1]

The OECD Investor Survey collected information on companies' characteristics for each survey respondent (e.g. sector and type); activities (e.g. involvement in international investment, operation and segment of the value chain); and type of policy measures that the investors had encountered, resulting in differentiated treatment vis-à-vis domestic firms. Key measures surveyed included:

- **Regulatory restrictions on FDI**, such as limits on foreign ownership;
- **Local-content requirements** (e.g. attached to a feed-in-tariff programme, a public tender or a financial scheme);
- **Differentiated or restricted access to financing** (e.g. capital grants or preferential loans);

- **Administrative and technical barriers** (e.g. burdensome permitting, licencing or certification procedures, or restricted access to the grid) or divergent technical standards; and
- **Trade measures**, other than local-content requirements, such as import tariffs, custom procedures and trade remedies (e.g. anti-dumping and countervailing duties).

As defined in this Annex, "international investors" refer to the survey respondents whose organisations had already invested or participated in solar-PV or wind-energy projects in foreign countries. "Domestic investors" refer to the respondents whose organisations had only invested or participated in solar-PV or wind-energy projects in their home countries.

The OECD Investor Survey asked each respondent to rank the impediment score of each policy restriction selected, from minor (1) to major (3), including medium (2). It also surveyed what international investor respondents perceived to be the impacts of each selected policy on their companies, in the concerned solar-PV and wind-energy markets. Respondents were presented with different statements to capture their views on the perceived impacts of each selected policy measure:

- The selected measure ("it") encouraged them to invest in local manufacturing or to source our inputs locally, or both;
- It discouraged them from investing in renewable-energy plants;
- It increased the costs of inputs for downstream power installations and related services;
- It reduced demand for renewable-energy-based power installations;
- It increased investment risk and uncertainty; and
- It had other effects.

Key Results

According to results of the 2014 OECD Investor Survey, LCRs were seen by international investors in solar and wind energy as the main policy restriction vis-à-vis foreign investors. Local-content requirements were identified by 71% of international investors as an impediment to international investment in solar PV and wind energy. Around 32% of international investors estimated that LCRs were a major impediment[2] (Figure 3.A2.1). Other measures considered in this survey included: administrative and technical barriers (42% of international investors identified it as an impediment); regulatory restrictions on FDI (36%); restricted access to financing (31%); and trade measures such as import tariffs, custom procedures and trade remedies (33%).

Figure 3.A2.1 **Policy measures identified by international investors as an impediment**[3]

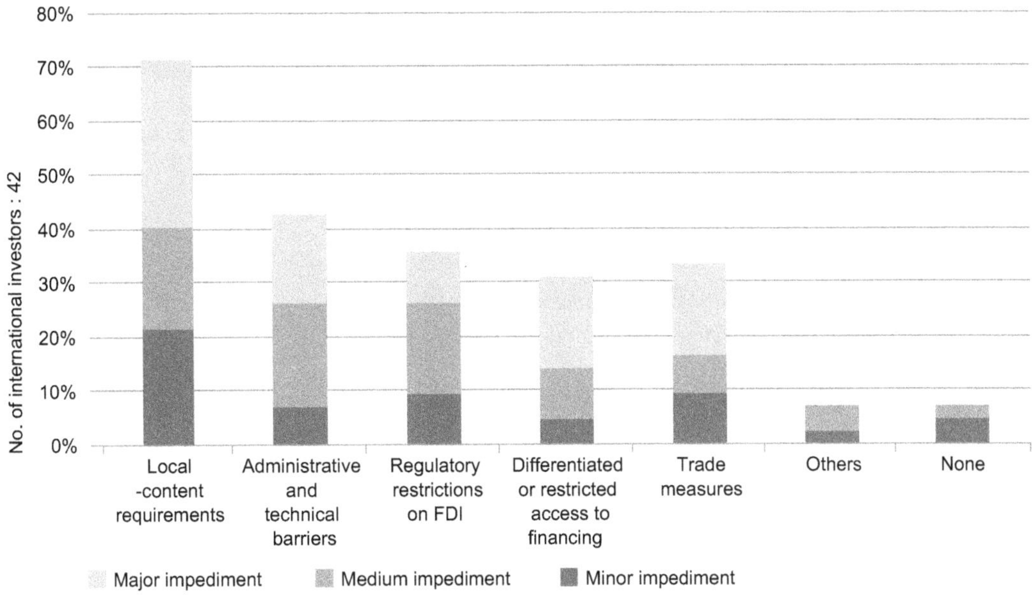

1. i.e. resulting in differentiated treatment for international firms vis-à-vis domestic firms.

Profile of the respondents

Project developers for solar- and wind-power plants accounted for the largest share (24%) of total respondents, including both international and domestic investors. They were followed by: service providers and financing institutions (16% respectively); equipment manufacturers (13%), banks (11%), and electric utilities (6%) (Figure 3.A2.2). Investors involved exclusively in downstream activities represented the majority of survey respondents (67%) (Table 3A2.1).

Figure 3.A2.2 **Type of companies surveyed in solar PV and wind energy** (in %)[4]

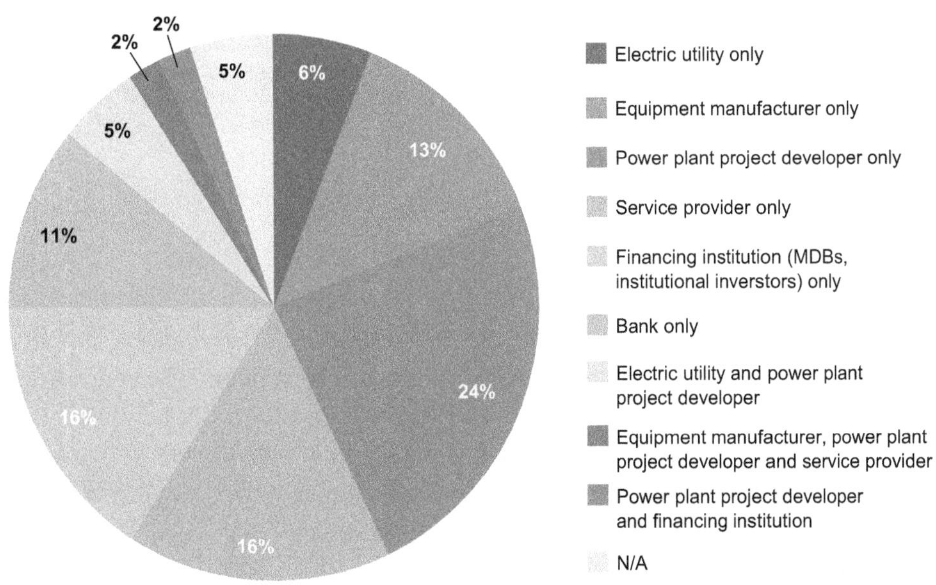

Table 3.A2.1 **Respondents from different segments of the solar-PV or wind-energy value chains** (in %)[5]

Segment	%
Upstream and midstream activities (manufacturing) only	14%
Downstream activities (e.g. power generation, project development, system integration, installation, maintenance, and financing) only	67%
Both	19%

A majority (68%) of the survey respondents were international investors, while 27% of them were domestic investors (Figure 3.A2.3). In addition, a majority (72%) of international investors operated in both solar PV and wind energy (Table 3.A2.2).

Figure 3.A2.3 **Percentage of international investors versus percentage of domestic investors**[6]

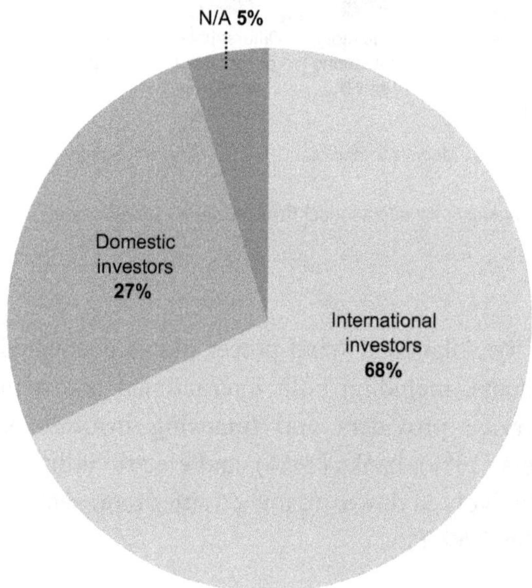

Table 3.A2.2 **Percentage of international investors involved in solar PV and wind energy**[7]

Sector	%
Solar-PV sector only	14%
Wind-energy sector only	14%
Both sectors	72%

LCRs as top impediments

According to results from the survey, LCRs were seen by international investors as the main policy restriction to their potential cross-border investments. The majority (71%) of international investors selected LCRs. This result is consistent with key findings from the report. Local-content requirements were seen as the top impediment by international investors involved in either midstream or downstream activities. About one-third of international investors identified LCRs as a major impediment and 21% identified it as a medium impediment (Figure 3.A2.1).

International investors demonstrated similar positions on LCRs whether they came from the solar-PV or wind-energy sectors (Table 3.A2.3). This result is unsurprising, given that most of the international investors surveyed (72%) operated both in solar PV and wind energy (Table 3.A2.2). Unsurprisingly, a majority of international investors involved in downstream activities of the solar and wind-energy value chains selected LCRs as an impediment (72% in wind energy and 73% in PV energy; Table 3.A2.3). What is less expected is that a majority of international investors involved in upstream and midstream activities also identified LCRs as an impediment (70% in wind energy and 75% in PV energy; Table 3.A2.3). This result may suggest that LCRs are seen as hindering international investment across the entire solar-PV and wind-energy value chains, not only in downstream activities. In addition, survey respondents reported that they encountered LCRs across 11 OECD countries, emerging economies and developing countries.

Table 3.A2.3 **Percentage of international investors from different segments of value chains and sectors who identified LCRs as a an impediment** (in %)

Solar-PV		Wind	
Upstream and midstream	Downstream	Upstream and midstream	Downstream
75%	73%	70%	72%

Note: Based on a sample of 42 international investor respondents. The results (in %) include responses from 8 international investors who operate across upstream midstream and downstream segments.

An important share of the surveyed international investors perceived that LCRs had negatively impacted their company's activities in relevant solar-PV and wind-energy markets. Key perceived impacts included: increased investment risk and uncertainty (48% across segments; and 64% in upstream or midstream segments); increased cost of intermediary inputs for downstream installations and related services (41% in the downstream segment); effect of discouraging international investors from investing in renewable-energy plants (46% in the downstream segment); and reduced demand for renewable-energy installations (10% across segments). Conversely, only 29% of international investors across the value chain segments responded that the presence of LCRs had encouraged them to invest in local manufacturing or to source their inputs locally.

In the open-ended questions of the survey, international investors' respondents stressed that LCRs distorted trade and adversely impacted their abilities to compete globally. Several project developers wrote that in several countries successful bidders for public tenders had included high shares of local content. They also noted that foreign banks were more likely to finance turbines that were manufactured in their respective home countries. Other international investors from financial institutions claimed that their institutions would not pursue any opportunity in a given country if the LCR policy in place imposes the local-production share that is higher than the one that they would expect without an LCR.

A few respondents expressed more positive views on the use of LCRs. Several project developers and equipment manufacturers said that LCRs made economic sense and could create local jobs in developing countries. However, they emphasised that LCRs should be used to support local job creations, rather than to promote a national industrial champion.

Other Measures

Administrative and technical barriers

Administrative and technical barriers were the second most significant policy restrictions according to the surveyed international investors. Around 17% of them selected these barriers as a major impediment (Figure 3.A2.1). In the open-ended questions, a few international investors from electric utilities and banks argued that different permitting procedures and grid-access regulations were major investment barriers, as well as frequent changes in those measures. They expressed frustration that their organisations faced such measures in all the countries in which they operate. The surveyed international investors listed 12 OECD, emerging and developing countries in which they faced administrative and technical barriers to international investment.

Regulatory restrictions on FDI

Regulatory restrictions on FDI were the third major policy restriction faced by international investors. Around 10% of the surveyed international investors identified FDI restrictions as a major impediment and 17% as a medium impediment (Figure 3.A2.1). Several project developers said that locals preferred working with firms with domestic ties. They said they faced the measures in 7 countries.

Trade measures such as import tariffs, custom procedures and trade remedies

Trade measures such as import tariffs, custom procedures and domestic trade remedies were identified as a major impediment by 17% of international investors (Figure 3.A2.1). A few international investors from financing institutions and banks argued that import duties on panels were particularly challenging.

Differentiated or restricted access to financing

Survey results also indicated that differentiated or restricted access to financing was another significant policy restriction for international investors. Around 17% of international investors said it was a major impediment and 10%, a medium impediment (Figure 3.A2.1). Several equipment manufacturers and project developers emphasised that imposing LCRs as a condition to benefit from preferential access to financing (e.g. by national banks) made their business unsustainable.

Results from the SETI Alliance survey

The Sustainable Energy Trade Initiative (SETI Alliance)[8] launched in 2014 a similar survey, in co-operation with the OECD. According to SETI Alliance results, LCRs constitute one of the main trade barriers for private sector stakeholders in sustainable-energy goods and services. More than two-thirds of respondents agreed that LCRs increased the cost of inputs for downstream power installations and related services. A similar proportion of respondents also agreed that LCRs reduced demand for renewable-power installations. Conversely, 80% of the respondents disagreed that LCRs encouraged them to invest in local manufacturing or to source out inputs locally.

Notes

1. The online questionnaire of the OECD Investor Survey is available at: http://webnet.oecd.org/Survey/Survey.aspx?s=1e130bc9e11a4f84ae078c07a523aa81.

2. The respondents who identified LCR as a policy impediment were asked to score the impediment as either: minor (1); medium (2); or major (3).

3. I.e. resulting in differentiated treatment for international firms vis-à-vis domestic firms.

4. Based on 62 respondents including both international and domestic investors.

5. Based on 42 respondents including only international investors.

6. Based on 62 respondents including both international and domestic investors.

7. Based on 42 respondents including only international investors.

8. SETI Alliance launched an on-line survey on its website http://seti-alliance.org/en/questionnaire in April 2014 to take stock of the key concerns of the private sector on trade in sustainable-energy goods and services. The survey results cited here are based on 30 responses that were gathered from April to July of 2014. The respondents are from private sector involved in renewable energy industry including solar PV, wind, biomass, energy-efficiency technologies, hydroelectric equipment and related services. The aim of the survey is to provide up-to-date information on the current state of play to the 14 governments engaged in multilateral negotiations on environmental goods.

References

Australian Productivity Commission (2010), "Bilateral and regional trade agreements", *Research Report*, No. 43, Productivity Commission, Canberra.

Bahar, H., J. Egeland and R. Steenblik (2013) "Domestic incentive measures with possible trade implications", *OECD Trade and Environment Working Papers*, No. 2013/01, OECD Publishing, Paris, http://dx.doi.org/10.1787/5k44srlksr6f-en.

Barker, M. (6 February 2013), "What's next for solar trade disputes in 2013?", Solarbuzz, http://www.solarbuzz.com/resources/blog/2013/02/what's-next-for-solar-trade-disputes-in-2013.

Bellmann, C. (2013), "Climate change and renewable energy development: The role of trade policy frameworks", *Multi-Year Expert Meeting on Commodities and Development 2013*, UNCTAD, Geneva.

BNEF (2013a), "New investment in clean energy fell 11% in 2012", Press Release of Bloomberg New Energy Finance.

BNEF (2013b), "2012 BNEF Clean energy & energy smart technologies league tables – The results", White Paper of Bloomberg New Energy Finance.

BNEF (2013c), *BNEF Database*, Bloomberg New Energy Finance.

Böhringer, C. et al. (2012), "Green jobs and renewable electricity policies: employment impacts of Ontario's feed-in tariff", *The B.E. Journal of Economic Analysis & Policy*, Vol. 12, Issue 1, Article 25, De Gruyter.

Cárdenas Rodríguez, M., et al. (2014), "Inducing Private Finance for Renewable Energy Projects: Evidence from Micro-Data", OECD Environment Working Papers, No. 67, OECD Publishing, Paris, http://dx.doi.org/10.1787/5jxvg0k6thr1-en.

Cimino C.(2013), "Local content requirements", *UNCTAD Ad Hoc Expert Group Meeting*, UNCTAD, Geneva.

Creed, A., and A. Kordvani (2013), *The WTO report on local content requirements for Canadian renewables programme - MENA implications*, Clyde & Co Insights.

EBRD (2012), "Local content requirements for renewable energy: An unnecessary evil", EBRD blog, European Bank for Reconstruction and Development, www.ebrdblog.com/wordpress/2012/11/local-content-requirements-for-renewable-energy-an-unnecessary-evil.

EWEA (2013), *Eastern winds: Emerging European wind power markets*, European Wind Energy Association, Brussels.

Frankfurt School-UNEP Centre, BNEF(2014), *Global Trends in Renewable Energy Investment 2014*, Frankfurt School of Finance & Management gGmbH.

Freshfields (2011), *Outlook for Infrastructure 2011: Getting Europe Back on Track – Will the 2020 Project Bond Initiative kick-start infrastructure investment?*, Freshfields Bruckhaus Deringer.

Ganesan K. et al. (2014), "Assessing Green Industrial Policy: The India experience", Research Report of Council on Energy, Environment & Water (CEEW), the International Institute for Sustainable Development.

Gaukrodger, D. and K. Gordon (2012), "Investor-state dispute settlement: A scoping paper for the investment policy community", *OECD Working Papers on International Investment*, 2012/03, OECD Publishing. Paris, http://dx.doi.org/10.1787/5k46b1r85j6f-en.

Ghosh, A. and R. Meléndez-Ortiz (2013), "Want clean energy? Avoid trade disputes", *Business Standard*, Business Standard Ltd.

Haščič, I., et al. (2015), "Public Interventions and Private Climate Finance Flows: Empirical Evidence from Renewable Energy Financing", OECD Environment Working Papers, No. 80, OECD Publishing, Paris, http://dx.doi.org/10.1787/5js6b1r9lfd4-en

Hufbauer, G.C. et al. (2013), *Local Content Requirements: A Global Problem*, Peterson Institute for International Economics, Washington, DC.

ICTSD (2013a), "Policy dialogue at the occasion of the Clean Energy Ministerial (CEM)", International Centre for Trade and Sustainable Development-CEEW (Council on Energy, Environment and Water), New Delhi, 16 April 2013.

ICTSD (2013b), *Local Content Requirements and the Green Economy,* International Centre for Trade and Sustainable Development, Geneva.

ICTSD (2013c), *Addressing Local Content Requirements in a Sustainable Energy Trade Agreement,* International Centre for Trade and Sustainable Development, Geneva.

ICTSD (2011a), *Fostering Low Carbon Growth: The Case for a Sustainable Energy Trade Agreement*, International Centre for Trade and Sustainable Development, Geneva.

ICTSD (2011b), *Feed-in Tariffs for Renewable Energy and WTO Subsidy Rules,* Trade and Sustainable Energy Series, Issue paper 4, International Centre for Trade and Sustainable Development, Geneva.

IEA (2014a), Energy Technology Perspectives 2014, IEA, Paris. http://dx.doi.org/10.1787/energy_tech-2014-en.

IEA (2014b), *Policy and Measures Databases* (database), www.iea.org/policiesandmeasures (accessed 27 August 2014).

IISD (2012), "Trends in investor claims over feed-in tariffs for renewable energy", interview with Brazilian Wind Energy Association (ABBeolica), EWEA, 10 March 2014.

IRENA (2013a), *Renewable Energy and Jobs* International Renewable Energy Agency, Abu Dhabi.

IRENA (2013b), *Renewable Energy Auctions in Developing Countries*, International Renewable Energy Agency, Abu Dhabi.

Johnson, O. (2013a), "Exploring the effectiveness of local content requirements in promoting solar PV manufacturing in India", Discussion Paper, 11/ 2013, DIE, Bonn.

Johnson, O (2013b), "When do local content requirements make sense? Lessons from India's national solar mission", presentation at UNCTAD Ad Hoc Expert Group on Local Content Requirements, Geneva, 13 June 2013.

Kuntze, J.C. and T. Moerenhout (2013), *Local Content Requirements and the Renewable Energy Industry – A Good Match?,* ICTSD, Geneva.

Lewis, J. et al. (2011), "Building a national wind turbine industry: experiences from China, India and South Korea", *International Journal of Technology and Globalisation*, Vol. 5, Nos. 3/4.

Morris, C. (2012), "German solar bubble? Look again!", *Series on the German Energy Transition,* No. 2, Heinrich Böll Stiftung (The Green Political Foundation), Washington, DC.

OECD (2015, forthcoming), “Emerging policy issues: Localisation barriers to trade, for an application of the trade model on quantitative restrictions to trade”, *OECD Trade Policy Working Paper*, OECD Publishing, Paris.

OECD (2015), *Policy Guidance for Investment in Clean Energy Infrastructure: Expanding Access to Clean Energy for Green Growth and Development*, OECD Publishing, Paris. http://dx.doi.org/10.1787/9789264212664-en.

OECD (2013a), *OECD Code of Liberalisation of Capital Movements,* OECD Publishing, Paris, http://www.oecd.org/daf/inv/investment-policy/CapitalMovements_WebEnglish.pdf.

OECD (2013b), *Inventory of Estimated Budgetary Support and Tax Expenditures for Fossil Fuels 2013*, OECD Publishing, Paris. http://dx.doi.org/10.1787/9789264187610-en

OECD (2013c), *Interconnected Economies: Benefiting from Global Value Chains*, OECD Publishing, Paris. http://dx.doi.org/10.1787/9789264189560-en.

OECD (2013d), *OECD Glossary of Statistical Terms*, http://stats.oecd.org/glossary/.

OECD (2013e), “OECD-WTO Database on Trade in Value-Added FAQs: Background Note”, OECD, http://www.oecd.org/sti/ind/TIVA_FAQ_Final.pdf.

OECD (2013f), “The Climate Challenge: Achieving Zero Emissions”, lecture by the OECD Secretary-General, Mr. Angel Gurría, London, 9 October 2013.

OECD (2013g), “OECD roundtable consultation on achieving a level playing field for international investment in solar and wind energy”, Paris, 6 December 2013, Summary report.

OECD (2012a), *Competitive Neutrality: Maintaining a Level Playing Field between Public and Private Business,* OECD Publishing, Paris.

OECD (2012b), *Attracting Investment in Renewable Energy in Ukraine*, Private Sector Development Policy Handbook, OECD Publishing, Paris.

OECD, WTO and UNCTAD (2014), “Report on G20 trade and investment measures (mid-November 2013 to mid-May 2014)”.

OECD, WTO and UNCTAD (2013a) “Implications of global value chains for trade, investment, development and jobs”, prepared for the G-20 Leaders Summit at Saint Petersburg, September 2013.

OECD, WTO and UNCTAD (2013b), “Report on G20 trade and investment measures (mid-October 2012 to mid-May 2013)”.

Rennkamp, B., and F. Fortes Westin (2013), *Feito no Brasil? Made in South Africa? Boosting technological development through local content policies in the wind energy industry,* Energy Research Centre (ERC), University of Cape Town, Cape Town.

Rivers, N.J. and Wigle, R. (2011), "Domestic content requirements and renewable energy legislation", working paper, http://ssrn.com/abstract=2129808.

Sahoo, A. and G. Shrimali (2013), "The effectiveness of domestic content criteria in India's Solar Mission", *Energy Policy,* Vol. 62, 1470-80, Elsevier, Amsterdam.

SETI Alliance (2014), "SETI Alliance launched global sustainable energy trade barrier stocktaking exercise" media advisory, 24 March 2014, SETI Alliance.

Shrimali, G. and A. Sahoo (2014), "Has India's Solar Mission increased the deployment of domestically produced solar modules?", *Energy Policy,* Vol. 69, 501-509, Elsevier, Amsterdam.

Stepp, M. and R.D. Atkinson (2012), "Green mercantilism: Threat to the clean energy economy", ITIF.

Stephenson, S. (2013), Addressing Local Content Requirements in a Sustainable Energy Trade Agreement"; International Centre for Trade and Sustainable Development, Geneva, Switzerland, www.ictsd.org

WTO (World Trade Organisation) (2014a), *WTO Dispute Settlement Gateway*.

WTO (2014b), "Joint statement regarding trade in environmental goods", Davos, 24 January 2014.

WTO (2014c), "Rules of origin", http://www.wto.org/english/tratop_e/roi_e/roi_e.htm.

WTO (2014d), *Agreement on Subsidies and Countervailing Measures* ("SCM Agreement").

WTO (1994), *Agreement on Technical Barriers to Trade* ("TBT Agreement").

Wu, M. and J. Salzman (2014), "The next generation of trade and environment conflicts: the rise of green industrial policy", *Northwestern University Law Review*, Vol. 108, No. 2, 410-474.

Xianqiang, Mao (2013), "Policy implications of the case of Chinese renewable energy industry", Ad hoc Expert Group Meeting on Domestic Requirements and Support Measures in Green Sectors: Economic and Environmental Effectiveness and Implications for Trade, Geneva, 13-14 June 2013.

Chapter 4

Other policy-related financial, trade and technical barriers to clean-energy investments

This chapter discusses a number of policies other than local-content requirements that may also hinder international investment in the solar-PV and wind-energy sectors. Three types of measures are discussed at length in view of their importance and the availability of extensive research: (i) domestic incentive measures, that may differentiate between domestic and foreign investors, such as preferential access to finance, export subsidies, preferential tax incentives, and government provision of subsidised inputs; (ii) trade remedies, such as countervailing and anti-dumping duties, which are permitted by WTO rules under specific circumstances; and (iii) technical regulations and standards. A number of other measures are also briefly discussed. These include: (i) applied import tariffs, (ii) regulatory restrictions on FDI, (iii) cumbersome administrative procedures; (iv) restricted access to the grid; (v) non-transparent procurement processes; and (vi) trade-related investment measures (TRIMS) other than LCRs. The chapter discusses the mechanisms by which these measures may adversely affect international trade and investment, and explains the implications for policy makers.

Although the main emphasis of the report is on local-content requirements (LCRs), this chapter takes stock of other policies that may also hinder international investment in the solar-PV and wind-energy sectors. Based on research and evidence gathered on a broad range of policies that can involve differentiated treatment between domestic and international investors in solar PV and wind energy, this chapter focuses on three types of measures that have increasingly been used by OECD countries and emerging economies in solar PV and wind energy. These measures include:

- **Domestic incentive measures** with possible implications for international trade and investment, such as **preferential access to financing** or subsidies that improve the export performance of solar-PV and wind-energy components;

- **Trade remedies** (e.g. countervailing and anti-dumping duties), which are permitted under specific circumstances under WTO rules to protect domestic producers from imports that allegedly benefited from actionable subsidies or were sold at less than fair market value (dumping). This chapter discusses the impacts of trade remedies and policy implications; and

- **Technical regulations and standards.** Trade barriers other than LCRs include tariffs and non-tariff measures (NTMs), such as technical measures, or "technical barriers to trade" (TBTs).

Other measures discussed less extensively in this chapter because they are little used in the solar-PV and wind-energy sectors or would deserve further research (e.g. grid access) include:

- Applied **import tariffs;**[1]

- **FDI regulatory restrictions;**

- **Cumbersome administrative procedures**;

- **Restricted access to the grid**;

- **Non-transparent procurement processes**; and

- **Trade-related investment measures (TRIMS) other than LCRs.**

- Other NTMs that are not reviewed in this report include: non-tariff surcharges and taxes; quotas; and inadequate protection of intellectual property rights;

Domestic incentive measures with possible implications for international trade and investment

This section takes stock of domestic incentive measures with possible implications for international trade and investment. It builds on existing OECD work (Bahar et al., 2013).

In the past decade, governments have provided substantial support to domestic PV and wind-turbine manufacturers. They have done so mostly through tax reductions (84 countries) and feed-in tariffs (FiTs; 71 countries; IRENA, 2013a). Domestic incentive measures vary in their design, their purpose and their incidence (Bahar et al., 2013; Ghosh et al., 2013).[2] They can be classified into two main categories:

- Demand-side (or demand-pull) policies affecting consumption, such as FiTs, renewable portfolio standards and carbon taxes; and
- Supply-side (or technology-push) policies benefitting production, e.g. R&D, direct investment and financial and fiscal incentives (Bahar et al., 2013; Guérin and Schiavo, 2011).

Most forms of government support for renewable energy indirectly affect trade and investment to the extent that they change relative market prices domestically or abroad. They are, however, not protectionist per se – i.e., they do not necessarily have adverse effects on international trade and investment (Bahar et al., 2013). Demand-side policies only distort trade when they are associated with trade barriers, e.g. whenever an LCR or an import tariff is associated with a FiT. Supply-side policies can help address market failures such as knowledge or R&D gaps, and thus improve domestic technologies' competitiveness vis-à-vis foreign technologies. Most incentive schemes can create trade distortions or even trade disputes (Bahar et al., 2013; Guérin and Schiavo, 2011).

Several types of domestic incentive measures can differentiate between domestic and international investors. They include:

- Preferential access to financing (e.g. direct financial transfers such as grants, equity injections and subsidised loans) favouring domestic manufacturers or associated with a LCR (Chapter 3);
- Export subsidies;
- Preferential tax incentives (e.g. producer tax incentives, production-based tax credits, and investment-based tax incentives); and
- Government provision of below-market-price inputs.

Under the WTO's Agreement on Subsidies and Countervailing Measures ("SCM Agreement"), several domestic incentive measures might be classified as *actionable* or even *prohibited subsidies.* A *subsidy* is defined by three basic elements. It has to involve: (1) a "financial contribution"; (2) by a government or any public body within the territory of a Member; (3) which confers a benefit"[3] (Article 1; WTO, 2014d). The SCM Agreement distinguishes two categories of subsidies: prohibited and actionable:

- *Prohibited subsidies* include subsidies contingent upon export performance (i.e. export subsidies) or on the use of domestic over imported goods (i.e. "local content subsidies"; Chapter 3). Prohibited subsidies are considered to be damaging to other countries and must be "withdrawn without delay" according to the WTO rules, as they directly affect trade and are the most trade-distorting subsidies; and
- *Actionable subsidies* include subsidies that are specifically provided to an enterprise or industry or group of enterprises or industries (Article 2, WTO, 2014d). Actionable subsidies are subject to challenge, either through the Dispute Settlement Mechanism

or countervailing action, in the event that they cause adverse effects to the interests of another WTO member country, as discussed subsequently (OECD, 2010e). For actionable subsidies, adverse effects and prejudice must be demonstrated empirically by the complainant. Unlike for prohibited subsidies, if an actionable subsidy is successfully challenged at the WTO, the violating country only needs to remove the adverse effects of the measure rather than the measure itself (WTO, 2014d). A support measure that is not limited to a specific firm or group of enterprises or industries (e.g. a FiT without a LCR) is not considered as an actionable subsidy.

Monitoring domestic incentive measures and determining whether specific subsidies violate WTO rules, however, is difficult (Bahar et al., 2013; Lincicome, 2012; WTO, 2014d). There are no officially adopted guidelines for calculating all types of energy subsidies, and no harmonised reporting mechanism, despite substantial reporting efforts by the OECD and the IEA (Ghosh and Gangania, 2012; OECD, 2013b; IEA, 2012d; Bahar et al., 2013). Nonetheless, clear rules exist in some cases for calculating and reporting the subsidies under the WTO. For production subsidies for instance, the WTO established a harmonised approach for subsidy notification under Article 25 of the SCM Agreement. Due to the difficulties in monitoring subsidies and the lack of common reporting standards for collecting energy subsidy-related information, inter-country comparisons may not be entirely reliable (Lincicome, 2012).

Preferential access to financing

Research suggests that preferential access to financing (e.g. through subsidised loans, loan guarantees and direct financial transfers) has frequently been used in emerging economies to support technology development and deployment in solar-PV and wind-turbine equipment. In China for example, state-owned or state-controlled banks have provided more than USD 40 billion in preferential loans and credit lines to Chinese solar manufacturers in 2010 (Table 4.1). Other studies estimate that Chinese state-owned banks have provided around USD 18 billion in subsidised loans to domestic solar manufacturers since 2007 (Bahar et al., 2013). China also offers tax holidays for companies that qualify as "new technology enterprises" and operate in Special Economic-development Zones (SEZs). Since 2008, support to solar manufacturers might have contributed to boosting Chinese PV exports, thereby leading to restructuring in other manufacturing markets such as Europe and the United States and ultimately to over-capacity throughout the sector. This situation has led several countries to impose countervailing duties on solar panels imported from China, in an alleged effort to provide relief from unfair trade practices that have injured their domestic solar manufacturers, as discussed subsequently.

Table 4.1 **Loans and other credit agreements involving Chinese banks to Chinese solar companies between 1 January 2010 and 28 September 2011**

Company	Amount (USD million)	Banks
China Sunenergy	160	China Development Bank
Daqo New Energy	154	Bank of China
Hanwa SolarOne	1 000	Bank of China
Hanwa SolarOne	885	Bank of Shanghai
JA Solar	4 400	China Development Bank
Jinko Solar	7 600	Bank of China
LDK Solar	8 900	China Development Bank
Suntech	7 330	China Development Bank
Trina Solar	4 400	China Development Bank
Yingli Green Energy	179	China Citic Bank, Bank of China
Yingli Green Energy	5 300	China Development Bank
Yingli Green Energy	144	Bank of Communications
Yingli Green Energy	257	Bank of Communications
Total	**40 709**	

Source: Mercom Capital Group.

Preferential access to financing for state-owned companies (SOEs) can also hinder market access and investment of foreign and domestic independent power producers (IPPs) in solar and wind energy. This is particularly challenging in countries with financial systems dominated by large state-owned banks (especially in infrastructure sectors), such as Brazil, China or Ukraine (Bahar et al., 2013; Ghosh et al., 2012; Guérin and Shiavo, 2011; KPMG, 2012). Preferential treatment of SOEs can lead to crowding-out of other investors and IPPs, which are more vulnerable to policy shifts and uncertainty in financial markets than SOEs. State-owned enterprises may benefit from preferential subsidised loans from governments or state-controlled financial institutions, even under liberalised electricity markets (Capobianco and Christiansen, 2011). When such preferential treatment benefits SOEs acting as foreign investors or traders, it can also result in competitive neutrality issues in international markets (Kowalski et al., 2013).

Developed countries have also used preferential access to financing through direct financial support or tax credits to support the domestic solar-PV and wind industries, often as part of post-crisis stimulus packages. In the United States for example, about USD 14 billion in federal loan guarantees have been provided between 2009 and 2011 to domestic solar projects as part of various stimulus packages (Bahar et al., 2013).

Financial contributions such as grants, loans, equity infusions, loan guarantees and fiscal incentives that result in adverse effect can qualify as actionable subsidies, under the WTO rules. Between 2008 and 2012, most complaints and cases of injurious subsidies in solar PV and wind energy have targeted the following measures: preferential access to financing; preferential tax credits; access to raw materials, components, energy and land at below-market prices; and support for R&D and technology development (Haley and Haley, 2013a).

Export subsidies and export credits

Several producers have complained against the alleged use of export subsidies, leading to trade investigations (Table 4.2; Stephenson, 2013; Ghosh et al., 2012). Export subsidies may take the form of direct grants or concessional loans. Since they are contingent on export performance, they are prohibited under WTO law (SCM Agreement, Article 3.1a). Not all types of export subsidies, however, are prohibited by the WTO. Export credit agencies are allowed to provide financial support for exports, provided that it complies with the OECD Arrangement on Officially Supported Export Credits[4] (Box 4.1).

Box 4.1 The role of export credit agencies in supporting the solar-PV and wind-energy sectors

Export promotion measures have been used by governments to encourage exports and sales by local companies, mainly in the form of export credit finance from public or semi-private finance institutions. Official export credits are provided through export credit agencies (ECAs), which provide direct loans or guarantees for facilitating exports. Export credit terms are regulated by the OECD Export Credit Guidelines, which allow favourable terms for renewable energy. Around USD 2.7 billion in export credits for renewable energy was reported in OECD countries in 2005-09. This amount remains small relative to export credits for fossil-fuel sources. In 2009, projects supporting renewable energy and cogeneration or district heating accounted for only 2.2% of total energy-related official export credits from OECD countries to developing countries (USD 0.7 billion out of USD 32 billion in 2009).

In the United States, the Export-Import Bank (Ex-Im Bank) has a mandate to increase support for US renewable energy, through providing financing mechanisms. They including: working capital loan guarantees; export-credit insurance; and financing to help foreign buyers purchase US goods and services. Ex-Im has granted loans to Indian solar developers to support the purchase of US solar goods (notably, thin-film panels). First Solar, the world's largest thin-film panel producer, received USD 583 million in loan guarantees for projects in Canada and India. Ex-Im increased its funding for renewable-energy exports to USD 720 million in 2012.

In China, both the Export Product Research and Development Fund and China Export Import Bank have offered export credits to local companies contingent on export performance, while public insurer Sinosure has offered export insurance at favourable terms.

Sources: Guerin and Shiavo (2011); Bahar et al. (2013); Ex-Im Bank (2011); Martin (2012).

Possible impacts of subsidies on international trade and investment

Trade-distorting subsidies may hamper international competition and investment in solar PV and wind energy in several ways:

- Both demand-side and supply-side incentives are likely to alter the equilibrium price in domestic markets, thereby affecting investment, production and consumption decisions; they can also induce knock-on effects in foreign markets (Bahar et al., 2013);
- Export subsidies distort international trade by altering the commercial value of solar-PV and wind-turbine inputs. They facilitate export of solar and wind-energy products and services by otherwise non-competitive domestic manufacturers (Bahar et al., 2013; OECD, 2010e);
- Preferential access to finance, such as capital grants and preferential loans, may lead to inefficiencies and over-capacity in solar-PV and wind manufacturing. This is because it can reduce the incentive for national manufacturers to undertake cost reduction through product innovation and rationalisation of production (Ghosh *et al.,* 2012; Lincicome, 2012; OECD, 2010e). Indeed, the use of subsidies to reduce the costs of domestic environmental industries or increase their revenues may reduce technological innovation provided by international competition; and
- Origin-based differences in treatment by national banks can raise barriers to entry for foreign investors in power generation and manufacturing, by reducing access to non-concessional and concessional finance such as preferential loans, loan guarantees and risk insurance. This is important because power generation is capital-intensive (Guerin and Shiavo, 2011).

Trade remedies

This section takes stock of the increasing use of trade remedies in solar PV and wind energy since 2011, to address unfair trade practices. The alleged use of trade-distorting subsidies has led to the escalation of WTO member investigations and trade remedies since 2010. These disputes have arisen despite repeated commitments from governments to avoid protectionism and trade distorting subsidies (Ghosh et al., 2012; Lincicome, 2012; Bahar et al., 2013). This section also assesses the effects of rising multilateral trade disputes on international trade and investment, and discusses policy implications. It builds on research by the Swedish National Board of Trade (Kasteng, 2014, 2013; Kommerskollegium, 2013a, b).

Background on trade remedies

WTO rules provide for three kinds of trade remedies (also known as trade defence instruments) to remedy "*injury*" caused to domestic industries by allegedly unfair trade practices (Box 4.2; Kasteng, 2013):

- Countervailing duties (CVDs) or anti-subsidy measures;
- Anti-dumping duties (ADs), targeting dumped imports; and

- Safeguard measures[5] to protect against sudden increases in imports.
- The WTO Agreement on Subsidies and Countervailing Measures covers the use of countervailing measures to offset injury caused by subsidised imports, in addition to covering the provision of subsidies (WTO, 2014d; Wu and Salzman, 2014). The WTO ASCM does provide for disciplines on the use of certain subsidies and provides some mechanisms by which to enforce these disciplines. Some of these mechanisms include multilateral dispute settlement or unilateral countervailing action, whenever an investigation (WTO, 2014d).

A countervailing duty (CVD) is an additional levy imposed on imported goods to offset the injurious effects to the domestic industry of subsidies provided to producers or exporters by the government of the exporting country (OECD, 2013e). WTO members may impose a countervailing duty whenever it is determined that there are subsidised imports, *injury* to a domestic industry, and a *causal link* between the subsidised imports and the injury (Box 4.2; WTO, 2014d).

An anti-dumping duty (AD) is an additional duty levied on imported goods to offset the injurious effects the dumped imports might cause to the domestic industry. Similar to a CVD, an AD may not be imposed by WTO members unless it is determined that there are dumped imports, *injury* to a domestic industry, and a *causal link* between the dumped imports and the injury (Box 4.2). Anti-dumping measures are addressed under the WTO Agreement on Implementation of Article VI of the 1994 General Agreement on Tariffs and Trade (GATT) (also known as the "Anti-Dumping Agreement"; GATT, 1994).[6]

The WTO has created the legal framework for the national trade remedy legislations. The investigations, however, are conducted by each country, according to the national legislation that has to be based on WTO rules. The European Commission for instance makes the trade remedy investigations for the 28 EU member states.

In addition, trade investigations often impact imports from the time they are initiated (and not from the time the countervailing or antidumping duties are actually imposed). This is due to the unpredictability of duties in terms of scope, level and imposition date (Kasteng, 2013, 2014; Kommerskollegium, 2013a, 2013b; Lucenti, 2003).

Assessing the definition of origin is complex in the context of trade remedy investigations (Kasteng, 2014; Chapter 2). As emphasised by former EU Trade Commissioner de Gucht: in an age of "complex supply chains", "a lot of our imports are inputs for manufacturing that takes place here" and "a significant share of the value of the finished goods we import has its origin in Europe: we all know the difference between *Made in China* and *Made by China*" (De Gucht, 2012). For example, the EU had to make a product-specific and legally binding amendment of its non-preferential rules of origin of solar panels in order to create legal certainty as to the conditions under which trade remedies would be imposed. This amendment was required because "the complexity of the production and assembly operations [of solar panels] might or might not confer origin" (Kasteng, 2014).

Box 4.2 How to define and determine adverse effect and injury?

Trade remedies. including anti-dumping and countervailing duty measures, are used to remedy *injury* caused to domestic industries by allegedly unfair trade practices that negatively affect employment, productivity, profit or market shares. There are three types of possible *adverse effects* of subsidies, which can lead to trade remedy action by WTO member countries:

- *Injury* to a domestic industry caused by subsidised imports in the territory of the complaining Member. This is the sole basis for countervailing action;
- *Serious prejudice,* which usually arises as a result of adverse effects (e.g., export displacement) in the market of the subsidising member country or in a third country market. Thus, unlike injury, it can serve as the basis for a complaint related to harm to a member country's export interests; and
- *Nullification* or *impairment* of benefits accruing under the GATT 1994, when subsidisation undercuts the improved market access presumed to flow from a bound tariff reduction.

Typically, the criteria for imposing countervailing or anti-dumping duty measures include evidence of dumping or subsidisation, material injury, and a causal link between the alleged dumping or subsidisation and the material injury. Under both the Anti-dumping agreement and the SCM agreements, material injury is defined as material injury itself, threat of material injury, or material retardation of the establishment of a domestic industry. A determination regarding material injury should be based on an objective examination based on positive evidence of the volume and price effects of dumped or subsidised imports and the consequent impact of dumped or subsidised imports on the domestic industry. A causal relationship between the dumped or subsidised imports and the injury to domestic industry is based on an examination of all relevant evidence and should include the examination of any other known factors, other than the dumped or subsidised imports, which may be injuring the domestic industry.

If a causal link is found between dumping or subsidisation and injury to the domestic industry, injury caused to the domestic industry may be remedied by the imposition of anti-dumping or countervailing duty measures. Certain trade remedy legislations (e.g. in the EU) allow for the imposition of anti-dumping or countervailing duty at a level lower than the margin of dumping or amount of subsidy, where the lesser duty would remove the injury to the domestic industry. Trade remedies target the specific products and firms that are the cause of the injury in the exporting countries concerned. Rates for ADs and CVDs are applied to individual exporters based on the calculated amounts by which those exporters have been found to be dumping or subsidised. Normally, a higher country-wide duty level is applied to firms that have not cooperated with the investigating authorities in the course of the trade remedy investigation.

Sources: WTO (2014d); Kasteng (2014), "Trade Remedies on Renewable: Intermediate and Long-Term Solutions", Presentation at UNCTAD Clean energy and Trade Ad hoc Expert Group 2 on "Trade Remedies in Green Sectors: The Case of Renewables", Geneva; Kasteng (2013), "Trade Remedies on Clean Energy: A New Trend in Need of Multilateral Initiatives", The E15 Initiative "Strengthening the multilateral trading system"; Kommerskollegium (2013b), "Targeting the Environment: Exploring a New Trend in the EU's Trade Defence Investigations", National Board of Trade, Stockholm.

Taking stock of the increasing use of trade remedies

Trade remedies have increasingly been used since 2011-12 by OECD countries – and recently increasingly by emerging economies (Table 4.2). Countries have done so to defend domestic manufacturers against the alleged distortions and unfair trade practices in solar and wind energy (including dumping and actionable subsidies; Kasteng, 2014, 2013; Kommerskollegium, 2013a, b; Cimino and Hufbauer, 2014; Ghosh et al, 2012; Lincicome, 2012). Since 2010, governments have launched 15 anti-dumping investigations and nine countervailing duty investigations, as well as imposed nine duty measures and seven countervailing duty measures on products associated with solar PV and wind energy, most of them since 2012 (Table 4.2; updated as of August 2014).

Trade remedies have typically been used to protect industries in OECD countries that have become subject to unfair competition in an attempt to create a level playing field between different markets. Former EU Trade Commissioner de Gucht stated in 2010 that, "[i]n the absence of international anti-competition rules and of other rules associated with well-functioning markets, trade defence instruments are the only possible means of protecting our industry against unfairly traded goods" (De Gucht, 2010).

Trade remedies are increasingly being used by emerging economies. Most trade investigations now originate in top solar and wind-energy markets: China, the EU and the United States.

Table 4.2 **Summary of recent trade remedies in solar and wind energy since 2009**[1]

Complainant Country	Targeted country	Measure	Range of duties and margins	Initiation of investigation	Measure(s) in force	Reference
The United States	China and Chinese Taipei	ADs	20.86%- 27.59% (on selected firms); 24.23% (all other firms) on solar panels imports from Chinese Taipei	2012	2014 (preliminary)	US Department of Commerce (DoC) (2014a)
		CVDs	18.56% - 35.21% (selected firms); 26.89% (all other firms) on solar panels imports from China	2012	2014 (preliminary)	US DoC (2014b)
	China	ADs	18.32% - 29.14% (selected firms) ; 249.96% (all other firms) on solar panels imports from China	2011	2012 (final)	US DoC (2014c)
			44.99 – 47.59% (selected firms); 70.63% (all other firms) on wind towers imports from China	2011	2012 (final)	US DoC (2014d)
		CVDs	14.78% – 15.97% (selected firms); 15.24% (all other firms) on solar panels imports from China	2011	2012 (final)	US DoC (2014c)
			21.86% -34.81% (selected firms); 28.34% (all other firms) on wind towers imports from China	2011	2012 (final)	US DoC (2014d)
	Viet Nam	ADs	51.50% (selected firm); 58.49% (all other firms) on wind towers imports from China	2011	2012 (final)	US DoC (2014d)

Table 4.2 **Summary of recent trade remedies in solar and wind energy since 2009**[1] *... continued*

Complainant Country	Targeted country	Measure	Range of duties and margins	Initiation of investigation	Measure(s) in force	Reference
China	US	ADs	53.30% -53.70% (selected firms) ; 57% (all other firms) on solar-grade polysilicon imports from the US	2012	2013 (final)	MOFCOM (2013), Announcement No. 48
		CVDs	0%-6.5% (selected firms) ; 6.5% (all other firms) on solar-grade polysilicon imports from the US	2012	2013 (final)	MOFCOM (2013), Announcement No. 63
	EU	ADs	42% (selected firms) ; 14.3% (all other firms) on solar-grade polysilicon imports from the EU	2012	2014 (final)	MOFCOM (2014), Announcement No. 25
		CVDs	1.2% on solar-grade polysilicon imports from the EU	2012	2014 (final)	MOFCOM (2014), Announcement No. 26
	Korea	ADs	2.4 - 48.7% (selected firms) ; 12.3% (all other firms) on solar-grade polysilicon imports from Korea	2012	2013 (final)	MOFCOM (2013), Announcement No. 48
The European Union (EU)	China	ADs	0.4%-36.1% (selected firms) ; 25% (other firms) on Chinese solar glass	2013	2014 (final)	EC(2014), No 470/2014
			27.3% - 64.9% (selected firms); 53.4% (all other firms) on solar panels imports from China	2012	2013 (final)	EC (2013) ; EC (2013), Decision 2013/423/EU; EC (2012) EMO/12/647
			13.8% (all firms) on Chinese glass fibre filaments imports (used for the production of wind turbine blades)	2009	2011 (final)	(EU) No 812/2010
		CVDs	3.2% - 16.7% (selected firms); 17.1% (all other firms) on Chinese solar glass	2013	2014 (final)	EC (2014), No 471/2014
			3.5% -11.5% (selected firms); 11.5% (all other firms) on Chinese solar panels	2012	2013 (final)	EC (2013), IP/13/769
			Investigation initiated on allegations that imports of certain filament glass fibre products, originating in China, are being subsidised and are thereby causing material injury to the Union industry.	2013	To be announced	EC (2013/C 362/05)
India	China, Chinese Taipei, the US and Malaysia	ADs	Indian Department of Commerce ruled in May 2014 in favour of ADs (5% to 110%) on panels from China, Chinese Taipei, the US and Malaysia. In Aug. 2014, the Indian Ministry of Finance rejected the recommendation.	2012	Not imposed	India Department of Commerce (2014)

1. Updated as of 26 August 2014. The methodology used consists in counting the number of cases of investigations or imposition of ADs and CVDs per each targeted country, in consistency with monitoring by the Swedish National Board of Trade. For duties and margins, ranges indicate duties for specific firms; duties and margins for "all other firms" are weighted average duty applied country wide. This report does not consider trade disputes linked to upstream production of raw material, such as silicon metal.

Sources: Compiled from databases of the U.S. Department of Commerce, European Commission, India Department of Commerce, Ministry of Commerce of People's Republic of China, Global Trade Alert and BNEF; Kommerskollegium (2013a, b); Cimino and Hufbauer (2014); Kasteng (2013).

Impacts of trade remedies

Trade remedies can help offset the impact of foreign incentive measures that are alleged to result in a material injury on part or whole of the domestic industry. Several studies, however, emphasise that trade remedies in the solar-PV and wind-energy sectors have had a number of negative impacts (Kasteng, 2013, 2014; Kommerskollegium, 2013b; Cimino and Hufbauer, 2014). These include reduced international trade volumes, increased prices and increased investment risk.

Impacts on international trade

The direct impact of trade remedies is to reduce trade volumes on the imports affected by the measures. According to a global survey by Hufbauer and Cimino (2014), the total reduction of trade as a result of AD and CVD measures is estimated to reach USD 13.6 billion annually, including about USD 8.5 billion for crystalline silicon PV cells (Box 4.3 and Table 4.3). According to the Swedish National Board of Trade, the value of renewable-energy imports affected by these measures amounts to EUR 14 billion in the EU alone. This is almost 75% of the total amount of all the EU's trade remedies in force today, making renewable energy the most affected sector in this regard (Kasteng, 2013). Out of the EUR 14 billion amount, the value of imports affected by trade remedies for solar panels from China alone amounts to EUR 11.5 billion, i.e. 82% of the total (Kasteng, 2013, 2014; Kommerskollegium, 2013b).

Impacts on input prices: the case of solar-PV energy

In a context of global value chains, AD and CVD investigations are likely to increase the costs of inputs for downstream producers by reducing imports of cheaper components. This is particularly true in the solar-PV sector. The price of PV modules is likely to increase, given the proliferation of ADs and CVDs, and the scale of solar-PV imports affected by trade remedies (Kasteng, 2013, 2014; Kommerskollegium, 2013 a, b; SolarServer, 2014a; Martin, 2014).

In the EU for instance, the imposition of trade remedies on imported renewable-energy products is likely to increase costs for EU producers, according to the Swedish National Board of Trade, (Kommerskollegium, 2013b). This is because EU producers are globalised, i.e. they have outsourced parts of their production processes to third countries (Kasteng, 2013).

In the United States, the rise of duties against China will likely lead to increased costs of US PV modules, according to IFS Technology, one of the largest solar-PV module manufacturers in the US (SolarServer, 2014a). Chinese suppliers currently supply between 50% and 60% of the installed PV capacity in the US, according to recent estimates (Gibson, 2014). Of the 57.8 GW of global PV module capacity, around 40.5 GW will be affected by the rulings (SolarServer, 2014a). The price of Chinese crystalline silicon (c-Si) PV modules has increased in the past year following recent increases in duties (Table 4.4). As a result, US PV module prices could increase to USD 0.75-0.80 per watt (SolarServer, 2014a). The lowest-priced modules available in the US market range around USD 0.62-0.65 per watt. Solar-PV module prices could increase to USD 0.75-0.80 per watt in the United States, depending on the outcome of the ongoing trade case between China and the Unites States (SolarServer, 2014a).

Box 4.3 Trade remedies targeting the renewable energy sector

The report "Trade Remedies Targeting the Renewable Energy Sector" takes stock of the increasing use of trade remedies in renewable energy worldwide, especially in solar-PV energy. The report recorded 41 recorded AD or CVD cases between 2009 and early 2014, including 18 AD or CVD cases in the solar sector, and 7 cases in wind energy.

The report also assesses the estimated impact of ADs and CVDs on worldwide renewable-energy trade flows. It estimates the total reduction of trade resulting from AD or CVDs to be about USD 13.6 billion annually. Of this total, the largest share belongs to crystalline silicon (c-Si) PV cells (USD 8.5 billion), followed by: biofuels (USD 3.6 billion); solar grade polysilicon (USD 953 million); wind turbines (USD 393 million); solar glass (USD 61 million); and glass fibre products (USD 24 million) (Table 4.3). Since trade remedy penalties are effective for 5 years, global trade losses are estimated to reach USD 68 billion over a five-year period.

Table 4.3 **Renewable energy products targeted in AD and CVD investigations**[1]

		Total trade affected		Estimated trade reduced		
Product	Number of AD/ CVD cases	Value (US$ millions)	% of total global trade	Value (US$ millions)	% of total global trade	Global trade in targeted renewable energy produts (US$ billions)[2]
Biofuels (biodiesel and bioethanol)	16	9,404	3.6	3,605	1.4	259
Solar energy						
Crystalline silicon photovoltaic cells & modules	11	19,230	21.7	8,549	9.6	89
Solar grade polysilicon	5	2,144	29.5	953	12.0	7
Solar glass	2	146	6.8	61	2.8	2
Wind energy						
Wind turbine blades (glass fiber products)	2	238	3.2	24	0.3	7
Wind turbines	5	804	8.3	393	4.1	10
Total	41	31,965	8.5	13,584	3.6	374

Note: Trade affected and trade reduced authors' calculations; global trade from World Bank WITS database and authors' calculations. Estimated trade reduced is calculated assuming a -1.0 elasticity of import demand for foreign goods and multiplying the total *ad valorem* duty imposed, or the sum of AD and CVD rates, by the average trade over the specific period.

Source: Cimino and Hufbauer (2014), "Trade Remedies Targeting the Renewable Energy Sector", paper presented at the *Green Economy and Trade. Ad Hoc Expert Group 2: Trade Remedies in Green Sectors: the Case of Renewables*, Geneva, 3-4 April 2014.

Table 4.4 **Price trends of crystalline silicon (c-Si) modules as of February 2014**

Origin	Price (€/ Wp)	Trend since Jan 2014	Trend since Jan 2013
Germany	0.70	+1.4%	-10.3%
Japan and Korea	0.69	-1.4%	-16.9%
China	0.59	+1.7%	**+11.3%**

Source: SolarServer (2014b), PVX spot market price index solar-PV modules.

Solar-PV thin-film module prices have already increased as a result of the recent trade disputes. Globally, the price of solar thin-film panels reached an 11-month high in June 2014, after US regulators set preliminary duties on standard polysilicon modules from China and Chinese Taipei (Martin, 2014).[7]

Possible impacts on investment risk perception and competitiveness of renewable energy vis-à-vis fossil fuels

The escalation of trade remedies may also hinder competitiveness and investment in the solar-PV and wind-energy sectors (Konings and Vandenbussche, 2010; Vandenbussche and Zanardi, 2010). Higher input costs can negatively affect the renewable energy value chains, especially in the solar sector, since the competitiveness of global production, supply and value chains depends on the continued development of specialisation, skills and innovation (Kasteng, 2013). Trade remedies may restrict competition and thus obstruct technical development and innovation in PV products (Yong, 2014). The rise of trade remedies in the EU for instance is likely to decrease the competitiveness of EU producers, in a context of global production, supply and value chains (Kommerskollegium, 2013b). This can be particularly challenging for manufacturers that rely on costly long-term supply agreements. In addition, the increase of solar-PV prices as a result of trade disputes is likely to impact electricity prices. Trade remedies might indeed affect the entire renewable-energy production process if they are imposed on intermediate products, thereby leading to reduced investment in solar and wind energy.

Trade disputes have also increased investors' perception of investment risk and uncertainty. The unpredictability of duties adds further uncertainty on investment decisions for foreign investors and downstream project developers. In a number of trade-remedy investigations, renewable-energy importers have also faced the possibility that the trade remedies might be imposed retroactively, which creates further uncertainty. Perceived instability may hinder investment and reduce incentives to invest in next-generation technologies (Wu and Salzman, 2014; Aggarwal and Evenett, 2012; CEPR, 2012).

The rise of trade remedies can therefore decrease the cost competitiveness of renewable energy compared with fossil-fuels alternatives, which may slow and hamper the transition away from fossil fuels to clean energy (Kasteng, 2013, 2014). Trade remedies may make renewable energy more expensive and less accessible for user industries, while also leading to further measures and countermeasures (Kasteng, 2013). Several syndicates and industry coalitions in Europe, China and the United States have warned against the risk of slower renewable-energy deployment, as a result of anti-dumping and countervailing actions (AFASE, 2013; Curtin, 2013; Kasteng, 2014; Kommerskollegium, 2013b).

Implications for policy makers

All WTO member countries with corresponding legislation are entitled to use trade remedies, and renewable-energy products constitute increasingly internationally traded goods. There is thus no legal obstacle for member countries to apply trade remedies to such products (Yong, 2014). Nevertheless, it is important to assess the impacts of trade remedies across the entire PV and wind-energy value chains before the trade remedy measures are imposed (including the risk of retaliation and escalation of trade disputes, and effects on competitiveness and international investment flows in both midstream and downstream activities). Trade remedies cannot be used as a "panacea" to protect domestic manufacturers.

Options to address trade remedies

Several policy options are available for governments to limit the use of trade remedies, as emphasised by the Swedish National Board of Trade. Options include limiting trade remedies:

- In level, e.g. by using the *lesser duty rule* (i.e. by imposing duties at a level lower than the margin of dumping but adequate to remove injury). This would ensure that remedies are not higher than needed to effectively remove the injury inflected, e.g. on the EU market (Kasteng, 2013c);
- In scope, e.g. by reducing the scope of trade remedies to specific product or import value; and
- In time, by introducing a time limitation for the trade remedy to be in place (Kasteng, 2013c).

Another option is for policy makers to align (or even replace) anti-dumping rules with competition rules (Kasteng, 2013; Kommerskollegium, 2013a). Policies aimed at protecting local industry could usefully aim at targeting truly harmful anti-competitive behaviour (Kommerskollegium, 2013a, b). For example, competition rules could only apply if a company has a market share of 40% or over compared with anti-dumping rules that would be instituted even if the practicing exporter has a market share of only 1%. Under this approach, higher thresholds would be set for companies deemed to have a dominant position and price undercutting. Competition rules can also help increasing the stringency of international rules on the trade and investment effects of SOE behaviour, among other options. This is relevant since SOEs provide an important share of trade- and investment-restrictive incentive measures such as preferential loans, especially in emerging economies (Kowalski. et al., 2013).

The case of the EU internal market shows that the removal of trade remedies does not necessarily lead to increased price undercutting or lost market shares (Kasteng, 2014; Kommerskollegium, 2013a). After the EU enlargement in 2004, ten Eastern and Central European accession countries were integrated into the EU15 and existing anti-dumping measures were terminated overnight against these countries. After the abolition of anti-dumping measures, the level of price undercutting decreased slightly and the market share of most products from the accession countries remained largely unchanged (Kasteng, 2014; Kommerskollegium, 2013a). The elimination of trade remedies among the member states was de facto replaced by competition rule in the EU, the European Free Trade Area (EFTA) and the European Economic Area (EEA) (as well as in certain bilateral trade agreements, such as Australia-New Zealand and Canada-Chile; Kasteng, 2014).

The escalation of trade remedies also calls for improved policy coherence. Governments' policies on trade remedies could usefully be aligned with other policy goals such as increasing the competitiveness of renewable energy. As emphasised by the Swedish National Board of Trade, the escalation of trade remedies in solar PV and wind energy could hinder international trade and investment in solar and wind energy, thereby hampering the competitiveness of renewable energy vis-à-vis fossil-fuel sources. Improved policy coherence would ensure that policy makers account for the full economic and environmental impacts of trade remedies. In the EU for instance, the impact of trade remedies on the EU's climate policy is currently not being taken into consideration in EU's trade remedy investigations (Kommerskollegium, 2013b). Policy alignment could for instance be facilitated by conducting regulatory impact assessments (RIA), which provide policy makers with an overall assessment of the likely gains and losses associated with planned policies.

The rise of trade remedies also calls for international co-operation. Multilateral efforts could usefully discuss the current provisions of the WTO Anti-Dumping Agreement (Article VI of the GATT) and the WTO SCM Agreement (Kasteng, 2013). In particular, it might be relevant for WTO member countries to consider the inclusion of environment-specific provisions with trade remedy agreements (Kasteng, 2013). Countries interested in further climate change and green growth co-operation could even agree on a possible Sustainable Energy Trade Agreement (SETA), to address renewable energy-related trade governance and the resulting legal challenges and opportunities (Stephenson, 2013). In January 2014, 15 countries[8] signed in Davos a *Joint Statement Regarding Trade in Environmental Goods* to announce their commitment to achieve global free trade in environmental goods, through reducing tariffs to 5% or less by 2015 on 54 environmental goods (WTO, 2014b; APEC, 2011).

Technical barriers to trade

This section takes stock of technical barriers in the wind-energy sector. Trade barriers other than local-content requirements include tariffs and non-tariff measures (NTMs). Non-tariff measures consist of technical barriers to trade (TBTs), such as: technical regulations; domestic voluntary product standards applied by the importing country that are not aligned with international standards and create regulatory divergence ("divergent national standards"); and certification and conformity assessment procedures (Steenblik et al., 2009; OECD, 2009; Steenblik and Matsuoka, 2009; Kirkegaard et al., 2010). New research by the European Wind Energy Association (EWEA) shows that technical barriers to trade have been used by countries to protect domestic manufacturers in wind energy. Such restrictions can hinder international investment in wind-power plants, e.g. by increasing transaction costs for foreign investors.

The WTO Agreement on Technical Barriers to Trade (TBTs) contains rules that set criteria used to determine whether technical regulations, standards, as well as testing and certification procedures, create non-tariff barriers and unnecessary obstacles to trade

(WTO, 1994). Technical regulations and voluntary standards require a product to fit specific characteristics (e.g. regarding its size, functions, performance, labelling or packaging) before entering a given marketplace. These measures usually serve legitimate goals of public policy, such as human health, safety or environmental protection. Nevertheless, they vary from country to country and can be more costly than necessary to achieve their desired objectives. They can also distort international trade and investment. According to the 11th report on *G20 Trade and Investment Measures*, a number of issues relating to technical barriers to trade for clean-energy products were raised at the April 2014 meeting of the Council for Trade in Goods. Those targeted primarily biofuels, which are not reviewed in this report (OECD, UNCTAD and WTO, 2014). On average, divergent national standards are relatively rare in the solar-PV industry, which is characterised by standardised technologies (Ghosh and Meléndez-Ortiz, 2013). In the wind-energy sector, by contrast, some countries have implemented nationally specific certification standards for wind-turbine components, which differ from internationally recognised standards,[9] as discussed below (EWEA, 2014; Annex 4.A; Lema et al., 2011; Stepp and Atkinson, 2012).

Technical restrictions to international trade and investment in wind energy

This section builds on analysis of technical restrictions to international trade and investment in wind energy in Brazil, Canada, China, India and South Africa, carried out with the support of the European Wind Energy Association (EWEA) (Annex 4.A; EWEA, 2014). It draws on various sources of information, including: surveys and interviews conducted by the EWEA with major wind-turbine manufacturers and developers, to cover at least 70% of the global wind energy market; the EWEA business intelligence database; desk research; resources from the national Wind Energy Associations; and information from international organisations and government institutions.[10] Factsheets for each country are provided in Annex 4.A.

The analysis considers technical restrictions to international trade and investment along the different stages of a wind-energy project lifecycle. According to the findings, **divergent national standards** constitute the main technical restriction specific to foreign investors in wind energy in the selected countries. Other technical barriers that can equally affect foreign and domestic investors include: national certification procedures that are not aligned with international ones; grid connection standards; and complex environmental permitting and environmental impact assessments (EIAs).

For Brazil, the analysis concludes that technical requirements for wind energy do not represent a problem for international investors, since Brazil applies international standards for wind turbines. The EWEA also identified a number of potential non-technical issues for Brazil that could hamper both domestic and international investment in wind energy, such as environmental regulations. The latter often differ from region to region and can add transaction costs and delays when developing wind power in a different region, especially for foreign investors.

In Canada, national wind-turbine standards and certification bodies are not fully aligned with international ones due to customisation of requirements to specific Canadian conditions (e.g. low temperature conditions). These requirements equally affect international and domestic investors. National standards developed by the Canadian Standards Association (CSA) differ from international IEC standards. The CSA standards, however, only serve as guidance since none of the provinces have adopted or enforced them. Thus, most manufacturers refer to the IEC Standards for certification. In addition, the Canadian regulatory framework for wind power is subject to approvals involving federal, provincial and municipal authorities, which can increase administrative barriers for both domestic and international investors.

In China, research indicates that the presence of divergent national standards and regulations on intellectual property rights can hamper investment in wind energy by foreign developers. Since 2011, China requires all wind-turbine producers to meet and attain national government standards and test certifications in order to be eligible for domestic wind projects. Only national certificates are accepted. Since Chinese standards differ from international standards, foreign manufacturers must adjust their turbine models to meet Chinese standards. China's standard setting process thus favours domestic wind turbine technology. It creates trade and investment barriers by restricting market entry for wind-turbine imports. In addition, government procurement favours products awarded with Chinese intellectual property certificates (e.g. a certificate for advanced wind turbine designs). Few non-Chinese companies have managed to participate and win wind-energy auctions. Nonetheless, no specific technical regulation has been found that explicitly discourages international players from participating in the bidding process.

Wind turbines installed in India have to comply with Indian national standards and certification procedures, which are aligned with international standards while being adapted to Indian geo-physical conditions. Major stakeholders have expressed an interest in establishing a certification facility in India. Testing services are market-driven, so developers often choose the Indian Centre for Wind Energy Technology (CWET) – the national certification body – over various international testing laboratories because of competitive pricing. The CWET works closely with international experts to align standards. Environmental permitting, however, is long and subject to multiple local authorities and communities, which adds transaction costs for international (and domestic) investors. Beyond technical barriers, the EWEA notes other impediments to wind-energy deployment, such as: access to the underdeveloped grid; difficulties in acquiring land for projects; challenges in getting permission to transport wind nacelles, towers and blades; poor transportation infrastructure; and LCRs (Chapter 3).

International wind-turbine standards apply to South Africa, and as such, there is no differentiated treatment associated with standards. However, there are strict environmental and grid connection regulations, although those requirements do not appear to intend to discourage foreign companies, which account for most investment flows. In particular, grid connection standards differ from international standards, and environmental requirements are very strict with respect to geophysical requirements such as water consumption. For instance, written confirmation of water allocation for all the water consumption needs of the project has to be provided, which enhance transaction costs by imposing lengthy delays to gather the documentation required to participate in bidding processes and certification procedures.

Possible impacts of divergent national standards and certification requirements

Divergent national standards and certification procedures requiring foreign investors and exporters to undergo additional testing and certification to demonstrate conformity with local norms can hamper international investment, in addition to directly distorting international trade:

- They can drive up costs and may act as a *de facto* market-access barrier. In particular, they can increase transaction costs and reduce transparency for foreign project developers and investors. This is especially true when technical restrictions are used as "red-tape" measures (EWEA, 2014; Ghosh and Meléndez-Ortiz, 2013). This is for instance the case in the wind-energy sector, or in a few OECD countries that apply specific standards to thin-film solar-PV technologies (EWEA, 2014; Stepp and Atkinson, 2012).

- By increasing costs, they can also reduce the "bankability" of projects, and prevent project developers from securing financing. This is particularly risky for projects with high up-front capital costs, such as wind-power plants; and

- They can constrain foreign firms providing downstream services, such as site assessment, financial due diligence or project development in the wind-power sector.

Overall, regulatory diversity increases the need for transparency. Policy implications include the need for greater international regulatory co-operation (e.g. to harmonise standards or increasingly, support mutual recognition and other approaches). More research is needed to assess the impact of divergent national standards on technology transfer.

Table 4.5 **Key technical restrictions identified by the EWEA**

	Brazil	Canada	China	India	South Africa
Summary of key findings	**No significant technical requirement differentiating international investors vis-à-vis domestic ones** National system currently adapting to international standards, with organisational gaps to be filled	**No significant technical requirement differentiating international investors vis-à-vis domestic ones** Complex approval processes involving federal authorities, affecting both international investors and domestic ones	**Presence of divergent national standards in wind energy:** All wind turbine producers required to meet local government standards and test certifications in order to be eligible for domestic wind projects. Only national certificates are accepted	**No significant technical requirement differentiating international investors vis-à-vis domestic ones** Wind-turbine producers required to comply with Indian standards and certifications, which are aligned with international standards but adapted to local geo-physical conditions	**No significant technical requirement differentiating international investors vis-à-vis domestic ones** Strict rules for grid connection, yet no attempt to discourage foreign investors, who account for most investments
Wind-turbine standards	**Technical requirements not necessarily favouring domestic investors** IEC[1] Standards are applied and accepted	**Technical requirements not necessarily favouring domestic investors** CSA[2] standards adopted but not required for approval. Most OEMs[3] referred to IEC standards, with minor deviations for certification purposes	**Technical requirements explicitly favouring domestic investors** No international certificates accepted. National Chinese Standards and certification differ from IEC Standards Purchases made by state-owned enterprises (who own the majority of installed capacity) favour domestic-patented products	**Technical requirements not necessarily favouring domestic investors** The certification scheme is aligned with international standards but adapted to specific local conditions. CWET[4] interacts closely with international research institutes	**Technical requirements not necessarily favouring domestic investors** IEC Standards are applied and accepted
Grid connection requirements	**Technical requirements not necessarily favouring domestic investors** Adopting international standards Obligation falling on developers	**Technical requirements not necessarily favouring domestic investors** Based on North American standards	**Technical requirements not necessarily favouring domestic investors** Chinese Standard for Technical Specifications for grid-connection design of large wind farm, differing from IEC Standard	**Technical requirements not necessarily favouring domestic investors** New forecasting requirements by Indian CERC[5] contested in court. On-going learning process	**Technical Requirements not necessarily favouring domestic investors** Stricter requirements differing from OECD ones (e.g. Special requirements for fault-ride-through)
Environ-mental Impact Assessments (EIAs)	**No technical Requirements** Conflicts of jurisdiction between environmental regulatory bodies, affecting both international and local players	**Technical requirements not necessarily favouring locals** Frequent and major changes in EIA process	**No specific technical restrictions with respect to EIAs for wind development that favours domestic investors vis-à-vis foreign investors**	**Burdensome permitting procedures** Permitting long and subject to intervention by multiple local authorities and communities	**Technical Requirements not necessarily favouring domestic investors** Stricter with respect to some requirements (e.g. water consumption needs of the project)

Source: EWEA; [1]International Electro technical Commission; [2] Canadian Standards Association;[3] Original Equipment Manufacturers; [4] The Indian Centre for Wind Energy Technology; [5] Central Electricity Regulatory Commission.

Import tariffs

Applied import tariffs[11] are relatively low in OECD and most large emerging non-OECD economies, and thus represent only a minor barrier to international trade and investment to date. In most developed countries and emerging economies, applied import tariffs on solar-PV and wind-turbine equipment fall within the range of zero to 10% (Table 4.6). The largest markets for solar PV and wind energy – China, the EU and the United States – have no import tariffs on solar components, while their tariffs on wind components range from 0 to 8% (Table 4.6). Tariffs are bounded in almost all countries. Bound and applied tariffs are on average higher in non-OECD than in OECD countries. In OECD countries, bound tariffs are set close to the level of applied tariff rates, whereas many non-OECD countries there is still "tariff water" between their applied rates and their bound rates. When such a gap exists, low applied tariffs could be raised any time to as high as the bound tariff (Bahar et al., 2013).

Evidence shows that higher import tariffs are sometimes used as part of the design of green industrial policies (ICTSD, 2009). A few emerging economies have recently increased their tariffs on renewable-energy equipment, e.g. Brazil on small- to medium-sized wind turbines. Both developed and developing countries, however, tend to remove those tariffs once their domestic renewable industries reach maturity (Stepp and Atkinson, 2012). For instance, China removed in 2010 its tariff on imported wind turbines (as did Korea in 2009 for import tariffs on solar panel pieces and wind turbines).

Possible impacts of import tariffs on international trade and investment

Several developing countries have used import tariffs to support domestic manufacturers; however, results to date seem to have been minor (OECD, 2013b). In addition, the increasing use of trade remedies may cancel some of the positive impacts of relatively low import tariffs in solar and wind energy, as discussed subsequently. Tariffs can also induce international "tariff-hopping" investment in local production capabilities, depending on the size and attractiveness of the domestic market to foreign investors (Wu and Salzman, 2013). This development strategy is not specific to clean energy and is thought to have had mixed results to date (OECD, 2013b). In the context of global value chains, the adverse effects of tariffs may be compounded due to the high intensity of international trade in intermediate inputs (Chapter 2). Import substitution through tariff policies may result in higher costs or lower quality inputs for locally sourced equipment, which may hinder the deployment of solar and wind technology, and thus international investment in electricity generation from renewable energy.

Table 4.6 **Average import tariffs of major importers on selected renewable-energy technologies and parts, as of September 2014**

Country	Hydraulic turbines				Wind-powered generating sets		Photovoltaic cells			
	Of a power not exceeding 1 000 kW (HS 841011)		Parts (HS 841090)		(HS 850231)		Whether or not assembled in modules or made-up into panels (HS 854140)		Parts (HS 854190)	
	MFN Applied	Bound	MFN Applied	Bound	MFN Applied	Bound	MFN Applied	Bound	MFN Applied	Bound
Australia	5%	15%	5%	15%	0-5%	0-10%	0%	0%	0%	0%
Canada	0-2%	6.1-9.7%	0-2%	0-9.7%	0%	6.2%	0%	0%	0%	0%
Chile	6%	23-25%	6%	25%	6%	25%	6%	25%	6%	25%
EU		4.5%		4.5%		2.70%		0%		0%
Iceland	0%	14-18%	0%	14%	0%	24%	0%	0%	0%	0%
Israel	0%	5%	0%	5%	0%	5-12%	0%	0%	0%	0%
Japan	0%	0%	0%	0%	0%	0%	0%	0%	0%	0%
Korea	0-8%	0-13%	0-8%	0-13%	8%	not bound	0%	0%	0%	0%
Mexico	0%	35%	0%	35%	0-15%	35%	0%	35%	0%	35%
New Zealand	0%	0%	0%	0%	0-5%	16.5%	0%	0%	0%	0%
Norway	0%	6%	0%	6%	0%	3%	0%	0%	0%	0%
Switzerland	CHF 12/100 gross kg		CHF 7-16/100 gross kg		CHF 11/100 gross kg		0%	0%	0%	0%
Turkey	4.5%	not bound	4.5%	not bound		10.8-14%	0%	0%	0%	0%
United States	3.8%	3.8%	3.8%	3.8%	2.5%	0-2.5%	0%	0%	0%	0%
Argentina	14%	35%	14%	35%	0%	35%	0-12%	35%	0%	35%
Brazil	14%	35%	14%	25-35%	0%	35%	0-12%	0-35%	0%	25%
China	10%	10%	6%	6%	8%	8%	0%	0%	0%	0%
Egypt	5%	20%	2%	20%	5%	10%	0%	0%	0%	0%
India	7.5%	25%	7.5%	25%	7.5%	25%	0%	0%	0%	0%
Indonesia	5%	30%	5%	30%	10%	40%	0-5%	40%	0%	0%
Malaysia	0%	5%	0%	not bound	0%	30%	0%	0%	0%	0%
Russia	15%	15%	7.5-15%	15%	0%	5%	0-3.3%	0%	3.3%	0%
South Africa	0%	0%	0%	0%	0%	20%	0%	10%	0%	10%

1. Note, for many countries, the MFN tariffs applied to least developed countries are lower or zero. In addition, tariffs applied on goods imported from partners in bilateral or regional free-trade agreements are also frequently zero.

2. Chile applies different tariff rates in the Arica and Magallanes regions in order to help isolated communities.

Sources Adapted from Bahar et al. (2013), using for MFN Applied tariffs: Market Access Database (http://madb.europa.eu/mkaccdb2/indexPubli.htm); and for bound tariffs: WTO. Updated as of 25 September 2014.

FDI regulatory restrictions

In most countries, regulatory restrictions on FDI in solar and wind energy are relatively low, below the economy-wide average level of restrictiveness (Golub et al., 2011; World Bank, 2010a; Table 4.7). In several countries however, the solar- and wind-energy sectors are not explicitly regulated by energy or investment legislation. It is therefore unclear whether the relatively low level of foreign ownership restrictions in wind and solar energy to date results from a willingness to promote open trade and investment policies, or from the fact that these sectors are new and have not yet been the subject of systematic attention on the part of the regulator (World Bank, 2010a). In addition, many restrictions to FDI in renewable energy are not specific to the renewable-energy sector, and apply more broadly to the energy sector. Foreign ownership is less restricted in power generation than in transmission and distribution.

Table 4.7 **FDI restrictions in renewable energy** (% of foreign ownership permitted)

Country	Sector	Type of investment	% of foreign ownership permitted in power generation	% of foreign ownership permitted in transmission and distribution
US	Wind and solar	Greenfield	100	100
		Cross-border M&A	100	
Austria	Wind and solar	Greenfield	100	49
		Cross-border M&A	49	
Greece	Wind and solar	Greenfield	0	0
		Cross-border M&A	0	
Turkey	Wind and solar	Greenfield	100	0
		Cross-border M&A	100	
China	Wind and solar	Greenfield	100	49
		Cross-border M&A	100	
Indonesia	Wind and solar	Greenfield	95	95
		Cross-border M&A	95	
Mexico	Wind and solar	Greenfield	0	0
		Cross-border M&A	0	
Morocco	Wind and solar	Greenfield	0	0
		Cross-border M&A	0	
Thailand	Wind and solar	Greenfield	49	49
		Cross-border M&A	49	
Vietnam	Wind and solar	Greenfield	100	0
		Cross-border M&A	100	
Philippines	Wind and solar	Greenfield	40	40
		Cross-border M&A	40	
Malaysia	Wind and solar	Greenfield	30	30
		Cross-border M&A	30	

Sources: Adapted from World Bank (2013) and Golub et al. (2011).

Limits on foreign ownership are the most obvious regulatory restrictions on FDI in solar and wind energy. This is largely due to the fact that the energy sector is considered as strategic. Limits on foreign ownership include:

- Restrictions with thresholds or outright prohibition of foreign ownership. In some countries these restrictions target renewable energy or the entire electricity sector, which is often placed under state ownership or control based on strategic considerations (OECD, 2011; Golub *et al.,* 2011; World Bank, 2013, 2010a; Table 4.7). Several OECD countries have foreign ownership restrictions on public electricity utilities;

- Requirements for foreign companies to form partnerships – mostly through joint-ventures – with domestic firms, as a prerequisite to invest in domestic wind- or solar-power plants, with minimum ownership requirements for local partners. This requirement exists in several emerging economies, though it is sometimes not explicit for clean-energy investment (ITA, 2010; BKPM, 2010); and

- Other restrictions, e.g. by prohibiting for foreign companies from holding a majority stake in Clean Development Mechanism (CDM) projects (Ellis and Kamel, 2007; EUCCC, 2010).

Restrictions on foreign ownership can also take place through admission, screening and approval procedures for inward FDI (OECD, 2003). Countries can, for instance, reject the purchase of specific solar or wind-energy companies by foreign investors or their domestic affiliates based on national security concerns (Kirkegaard et al., 2010; OECD, 2008a). Restrictions of ownership and acquisition of land by foreign companies and non-citizens for business purposes can also limit FDI, as included in the OECD FDI restrictiveness Index (Kalinova et al., 2010).

Other restrictions to FDI are not considered here. They include: equity restrictions; constraints on foreign personnel and operational freedom; nationality requirements of board members; provisions regulating residence and issuance of work permits to foreigners; capital control limiting the repatriation of profits; and competition limits related to the existence of a public monopoly provider (OECD, 2012a, 2003; Golub et al, 2011; Ellis and Kamel, 2007). Barriers to ownership and acquisition of land by foreign companies can be particularly challenging in solar and wind energy, and would deserve further consideration. In Brazil for instance, foreign companies are prohibited from buying more than 250 to 3500 hectares (depending on the region), and more than 25% of municipal land (Annex 4.A).

Other restrictions

Administrative barriers

Administrative hurdles can restrict foreign companies' access to domestic wind and solar energy markets, even in the absence of restrictions to entry and establishment of majority foreign-owned businesses. Administrative procedures are often hard to navigate for foreign investors:

- They can involve differentiated treatment for foreign investors, e.g. with special permitting, licensing and certification procedures. Preferential administrative treatment for local companies can hinder the implementation of foreign investment projects; and

- They can generate higher transaction costs of establishment for foreign investors than for domestic ones, even in the absence of deliberate intent to raise administrative and technical barriers to entry for international investors. This is due to the lack of information and expertise of foreign entrants. Higher transaction costs can be generated by licensing, certification and permitting procedures to build and operate wind and solar-energy facilities, particularly in the absence of investment promotion agencies (IPAs) and one-stop shops, or in decentralised countries, where sub-national government can impose additional regulations (OECD, 2013a). For example, delays in construction permits can increase transaction costs and risks for international investors.

Restrictions on operations: the case of non-transparent government-procurement procedures

Obstacles to operations such as differentiated treatment or lack of transparency in government-procurement procedures can also restrict international investment in wind and solar energy. The OECD has conducted relevant work on this issue (Capobianco and Christiansen, 2011; OECD, 2013a; 2012a; 2010a, 2010b, 2005, 2004, 2002; Steenblik and Matsuoka, 2009). The absence of transparent, open and competitive bidding processes can constrain operational freedom of foreign investors in domestic wind and solar-energy markets, and hamper their ability to participate in bidding processes and concession tenders. This is sometimes the case for procurement of national concession projects in wind or solar energy (EUCCC, 2010). Public procurement procedures can raise concerns in countries with large state-owned enterprises (SOEs) in renewable energy (Capobianco and Christiansen, 2011; EUCCC, 2010). Even under concession models designed to promote competition, SOEs sometimes benefit from preferential treatment. For example, such models may overweight the value of low electricity tariffs in bidding evaluation criteria; this allows domestic SOEs to win bids because they face less pressure to charge enough to yield a market rate of return. Public authorities may also support SOEs by reserving a share of the planned new renewable generation capacity for them (OECD, 2013a). Additional obstacles arising during wind and solar project development for foreign investors can increase the operational risk of such projects and result in projects missing deadlines.

Other possible operational restrictions include asset repatriation, remittance restrictions and constraints on foreign personnel. Several foreign companies, for example, have complained about restrictive foreign exchange regulations and impediments to the repatriation of profits, royalties and other fees to home countries.

Restrictive access to the grid

Access to the transmission grid, network pricing and connection costs can also restrict entry for both foreign and domestic independent power producers (IPPs) in the power-generation sector (OECD, 2013a). Even after the establishment of a competitive electricity market, some state-owned solar and wind power plants continue to be provided with easier access to transmission by grid controllers than non-state investors in some countries. Lead times in obtaining existing grid connections or the development of new grid infrastructure are common barriers both for domestic and foreign IPPs and can delay projects and raise transaction costs. In addition, differentiated connection permission procedures can deter foreign investors from entering markets, by delaying projects or raising transaction costs. The OECD Corporate Governance Committee has undertaken relevant work on competitive neutrality issues and trade effects related to state-owned enterprises (Kowalski et al., 2013; Capobianco and Christiansen, 2011; OECD, 2005). More research is needed to further assess restrictions for IPPs in solar PV and wind energy in terms of access to the grid.

Trade related investment measures (TRIMS) and other measures affecting services

Trade related investment measures (TRIMS) other than local-content requirements include: local-equity requirements; technology transfer requirements; licensing requirements; and remittance restrictions. In particular:

- Local-equity requirements are often characterised by restrictions on registered capital in renewable-energy projects that are different for foreign and domestic investors. Some emerging economies impose a minimum share of registered capital for foreign and joint-venture invested projects (EUCCC, 2010). Uruguay, which has also auctioned new renewable-energy projects, includes stipulations in its tenders that foreign projects benefitting from its FiT must include a minimum of 20% equity participation by a local partner, and contract 80% of subsequent maintenance locally (Bahar et al., 2013).

- Technology transfer requirements can also be challenging for international investors in wind and solar energy. Such measures require foreign firms trying to enter the domestic market to relocate their R&D facilities or to transfer their technologies to domestic firms by imposing partnership and joint-venture requirements (Stepp and Atkinson, 2012). They are frequently used in the wind-power industry (e.g. in Portugal) and deserve further analysis. They can also have a positive impact by encouraging foreign investment, including when foreign companies receive preferential treatment in markets with low technological-development capacity, in exchange for establishing joint ventures and transferring technology at a lower cost (Bahar et al., 2013).

Investments in solar and wind energy services are not covered by the WTO Agreement on TRIMS, but by the General Agreement on Trade in Services (GATS), which is not discussed in the present report. Further research could usefully further assess the impact of trade- and investment-restrictive measures on downstream services in solar PV and wind energy.

Notes

1. There are two types of tariffs: bound and applied tariffs. Bound tariffs refer to the legally bound commitments on customs duty rates, which act as ceilings on the tariffs that member governments can set. Applied tariffs refer to the rates that governments actually charge on imports, which can be lower than bound tariffs and have a direct impact on trade.

2. For a typology of domestic incentive measures, see Figures 1 and 2 in Bahar et al. (2013).

3. I.e. if the government's contribution is more favourable than what would have normally been available for the renewable energy project developer or manufacturer on the open market; WTO (2013b).

4. The OECD has developed an agreed "Arrangement on Officially Supported Export Credits" to address climate change concerns. OECD countries have recently agreed on new rules to: (i) strengthen environmental and social due diligence processes of officially-supported export credits; (ii) create financially prudent incentives to support business projects with low-carbon emissions; and (iii) encourage support for advanced climate-friendly technologies. Under the 2012 Sector Understanding on Export Credits for Renewable Energy, Climate Change Mitigation and Water Projects, solar-PV and wind-energy projects may qualify for the financial terms applicable to officially-supported export credits for renewable energy projects.

5. Under the Agreement on Safeguards (Article XIX, GATT, 1994), safeguard measures are defined as "emergency" actions with respect to increased imports of particular products, where such imports have caused or threaten to cause serious injury to the importing Member's domestic industry. Such measures, which take the form of suspension of concessions or obligations, can consist of quantitative import restrictions or duty increases higher than bound rates.

6. The Anti-dumping agreement allows for the imposition of anti-dumping duties, as part of investigations conducted in accordance with the provisions of the Agreement, generally equal to the difference between the export price of the product exported from one country to another and the normal value (the "dumping margin"; OECD, 2013e).

7. Spot prices for thin-film modules increased 5.2% in June 2014 to 61.2 cents/KW, the highest since July 2013, while during the same period, the more commonly used polysilicon panels produced mostly in China declined 2.8%; Martin (2014).

8. Including: Australia; Canada; China; Costa Rica; the European Union; Hong Kong, China; Japan; Korea; New Zealand; Norway; Singapore; Switzerland; Chinese Taipei; and the United States; WTO, 2014b.

9. E.g. the International Electrotechnical Commission (IEC) and the International Organization for Standardization (ISO).

10. Such as the EU Directorate General for Trade (DG Trade), the *International Centre for Trade and Sustainable Development* (ICTSD), the United Nations Conference on Trade and Development (UNCTAD) and the WTO.

11. There are two types of tariffs: bound and applied tariffs. Bound tariffs refer to the legally bound commitments customs duty rates, which act as ceilings on the tariffs that member governments can set. Applied tariffs refer to the rates that governments actually charge on imports, which can be lower than bound tariffs and have a direct impact on trade.

Annex 4.A

Technical restrictions to international investment in wind energy in selected OECD and emerging economies: Country factsheets

Country factsheet of Brazil

Summary of findings

The analysis finds that Brazilian technical requirements for wind energy do not represent a problem for international investors, as the country conforms to international standards for wind turbines. Nonetheless, a number of potential non-technical issues could equally hamper both international and domestic investors in wind energy. They include: restrictions on land ownership for foreign investors; and conflicting jurisdictions between environmental regulatory bodies. Conflicting jurisdictions could result in delays and additional transaction costs for project developers and investors participating in wind-energy bids. Environmental regulations often differ from region to region and may require additional administrative work when developing wind power in a different region.[1]

Market Outlook

Brazil is the fastest-growing South American market. The country's wind resources represent a stable source of energy, with capacity factors twice as high as the European average. In 2013, Brazil installed 1.5 GW of new wind-energy capacity, of which one third as grid connected, bringing total wind power installations to 3.4 GW. The Brazilian government's Decennial Energy Plan set a goal of 17 GW of installed wind capacity by 2022, accounting for 9.5% of national electricity consumption (GWEC, 2012a).

Wind turbine standards

International standards apply to Brazilian wind turbines

Brazil's national system is currently adapting to international standards. There are three standards published by Brazilian National Standards Organisation (BNSO) relating to wind turbines in Brazil. These national standards are equivalent to International Electrotechnical Commission (IEC) standards. In the absence of BNSO standards, wind-turbine manufacturers can get their products certified through international standards (IEC standards) (Interview by EWEA with COBEI, 2014).

Grid connection requirements

The connection of wind farms to the grid is assessed on a project basis. Project developers carry the cost of identifying connection points to the grid. Since 2013, project developers also bear the transmission risk. It means that if the wind farm is ready before the connection, then the developer is obliged to buy electricity to fulfil his contract (Ministry of Mines and Energy, 2014).

Environmental Impact Assessment (EIA)

Environmental regulations often differ from state to state and from region to region. This requires additional administrative work for both international investors and domestic investors when developing wind power in a different region (Interview with Brazilian Wind Energy Association, 2014). There is no national regulation on EIA, and each state has its own laws due to environmental specificities relating to flora and fauna. Developers are obliged to meet the requirements of the environmental agency of the state in which they want to produce the respective environmental studies. For example, state legislations differ on:

- The required distance between wind turbines and inhabited communities;
- The types of studies required for obtaining licenses; and
- The size classification of wind farms.

Non-technical issues that could potentially hinder both international and domestic investors

Conflict of jurisdiction between environmental licence providers

The conflict of jurisdiction between different bodies when issuing environmental licenses is an important barrier faced by wind-energy companies in Brazil, although this does not treat domestic and foreign companies differently. Most of the potential land areas for wind-energy deployment are found in permanent preservation areas (APPs), which are subject to special protection measures, environmental impact assessments (EIAs) and licensing requirements (CERDI, 2010). State agencies are typically responsible for the licensing of wind-power plants in APPs, requiring a Simplified Environmental Report (RAS). When the projects are located in coastal areas, the competency for issuing the environmental license falls to federal agencies. Conflicts of jurisdiction can arise when the projects fall simultaneously in both APPs and coastal areas.

Land-purchase restrictions for foreigners

Land-purchase restrictions for foreign companies were introduced in August 2010. Foreign companies, even if acting through a subsidiary in Brazil, cannot buy more than 50 modules of land, varying between 250 hectares and 3500 hectares, depending on the region. Moreover, according to the regulation, no more than 25% of the land in a given municipality can be acquired by companies under foreign control (Latin American Herald Tribune, 2014). The legislation does not apply retroactively to existing properties.

Country factsheet of Canada

Summary of findings

The analysis finds that national standards and certification bodies in Canada are not fully aligned with international ones, due to customisation of requirements to specific Canadian conditions (e.g. low temperature conditions). This affects international and local investors equally. National standards developed by the Canadian Standards Association (CSA) differ from international IEC standards. The CSA standards, however, only serve as guidance since none of the provinces have adopted or enforced them. Thus, most manufacturers refer to the IEC Standards for certification. In addition, the Canadian regulatory framework for wind power is subject to approvals involving federal, provincial and municipal authorities, which can increase administrative barriers for both domestic and international investors.[2]

Market outlook

According to the Canadian Wind Energy Association (CAWEA), wind energy is expected to power 20% of the country's energy needs by 2025 (GWEC 2014). In 2013 alone, 1 599 MW of wind power was installed in Canada, bringing the total installed capacity in Canada to 7 803 MW. There are currently 185 wind farms installed in Canada, spread over 10 provinces and two territories. The majority of wind farms are located in Alberta, British Columbia, Ontario and Quebec.[3]

Renewable-energy regulations

Currently there is no federal legislation that supports renewable energy. Rules and regulations from provinces and territories are in place to develop renewable-energy projects. Only one province, Nova Scotia, has renewable-energy targets. The province of Ontario has a regulatory framework on the energy-supply mix, when renewable energy sources are included (CAWEA, 2014; Christopher et al., 2011).[4]

Administrative procedures for wind development

Project approvals set by national and provincial regulations

Wind-energy developers are subject to complex approval processes involving federal, state or municipal authorities. The environment is an area of shared jurisdiction, so all levels of government (federal, provincial, territorial and municipal) in Canada are involved to develop a renewable-energy project. Provinces have authority (along with the federal government and its National Energy Board) over energy policy. Not all projects are subject to federal oversight but there are federal permits related to species at risk, air traffic, etc. This can be further complicated by the jurisdictional makeup and ownership of the Canadian electrical transmission and distribution systems, which range from locally-owned private businesses to fully integrated provincially-owned systems, with a wide range of regulatory roles and connection requirements. This process affects all investors equally.

Wind turbine standards

National certification body

The Canadian Standards Association (CSA) has adopted the IEC standards with minor deviations. The CSA sets national standards for wind turbines that differ from the IEC standards, but without being enforced for certification. While the CSA standards continue not to be adopted, most manufacturers or "original equipment manufacturers" (OEMs) use the IEC standards for certification (CAWEA, 2014). In absence of specific and up-to-date regulations on wind energy, Canadian authorities having jurisdiction (AHJ) have developed a range of standard references on the characteristics of wind turbines, in order to evaluate and approve the effects of wind-turbine installations on the environment, zoning, power quality, grid integration, performance testing and safety. Standard references for evaluating wind-turbine projects are a combination of European standards and Canadian-specific standards. The CSA has participated in this effort since the 1980s, when the National Standards of Canada pertaining to wind turbines were developed. The CSA published guidelines on Canadian wind turbine codes and standards in 2008 which was updated in 2012 to reflect the current state of requirements within Canada.

Requirements relating to Canadian weather conditions

The current National Standard of Canada for wind turbines (CAN/CSA-F416) includes the design considerations for various environmental conditions, including hail, icing or seismic loads protection. Indeed, IEC standards (such as the IEC 61400 series) do not address extremely cold weather considerations, which are prevalent throughout much of Canada.

Electrical safety

Grid connected equipment must meet electrical safety requirements of the applicable AHJ, which may include the following organisations: the grid owner, the system operator and the provincial electrical safety authorities. Developers are required to contact the appropriate AHJ to confirm specific local requirements. There is some harmonisation between Canadian standards and IEC or European standards for electrical safety. While international companies are able to adapt their products to different standards, potential harmonisation would ease clarification of the possible acceptable solutions for electrical safety.

Grid connection requirements

Developers have to adapt to the respective electrical design requirements on a site-specific basis, since there is no grid code regulating connections at the national level. There is no indication of preference for local players over international ones. Most of the connection and system operation requirements, however, are clearly favouring hydroelectric power and not variable renewable energy sources such as wind energy. Power quality issues are addressed in North America by appropriate Institute of Electrical and Electronics Engineers (IEEE) and North American Electricity Reliability Corporation (NERC) standards which should be applied to wind farms, while IEC standards tend to be more prescriptive and definitive but are not used frequently in North America. Most Canadian provinces must follow NERC[5] standards, either through the utilities under provincial regulation or in provincial legislation. Virtually all provinces are a party to NERC and therefore must adopt NERC standards, though this is not always compulsory.

Environmental impact assessments

Each of the ten provinces and three territories has their own EIA procedures regarding renewable-energy projects and there are frequent major changes in these processes, affecting both international investors and local ones. Several provinces have been cooperating with the federal government in order to streamline the process and avoid duplicating efforts. The EIA can indirectly hinder developers, for example by increasing transaction costs, or for carrying out impact assessments on bats (CAWEA, 2014).

Country factsheet of China

Summary of findings

The analysis stresses that China has divergent national standards in wind energy, which can hamper international investors. China requires all wind-turbine producers to meet and attain local government standards and test certifications in order to be eligible for domestic wind projects. Only national certificates are accepted. Since Chinese standards differ from international ones, foreign turbine manufacturers must adjust their turbine models to meet Chinese standards. Additionally, obtaining the Chinese intellectual property certificate associated with advanced wind turbine designs is a key priority for the Chinese government, as it is developing policies that explicitly support Chinese wind turbine manufacturers. Products made with Chinese intellectual property qualify for priority in government procurement.[6] In practice, this means that international investors are generally unable to participate in government procurement.

Market outlook

China is the largest wind power market globally, adding more capacity each year than any other country. In 2013, 16.1 GW were installed, bringing the cumulative total installed capacity in wind energy to 91.4 GW. China's wind power potential is enormous. The commercial onshore potential is estimated between 1 000 and 4 000 GW, with an additional 500 GW of offshore potential. China aims to reach a minimum of 150 GW from wind power by 2020. The National Energy Administration's (NEA) 12th Five-Year plan for renewable energy for 2011-15 has a target of 100 GW of wind by 2015.

Administrative procedures for wind development

Only few non-Chinese companies have managed to participate and win Chinese wind energy auctions. However, no specific technical regulation has been found to be explicitly discouraging international players from participating in the bidding process.

Wind farm standards

Chinese national and professional standards

Although the accreditations of international standards compliance for the certification of wind turbines and general quality systems are recognised in China, the rapid development of China's wind power industry has also led to development of local standards. China's new national standards reflect its particular topographical and meteorological conditions (i.e. wind speed, low temperatures, high altitudes and typhoon conditions), which are not covered by IEC standards. There are two levels of standardisation that differ from international standards: national standards and professional standards. The NB[7] standards are the Chinese professional standards for industries. Commonly referred to as industry standards or sector standards, they are developed and applied when no national standard exists. In 2010, the National Energy Bureau (NEB) created under the National Development and Reform Commission (NDRC) issued 18 wind-power technical standards drafted by the Wind Power Standardization Technical Committee that covers wind power grid-connecting, project estimates, and generator unit equipment, which further improved and supplemented Chinese technical standards in the areas of wind-turbine generator system manufacturing, project construction costs, quality assurance, installation and operation and maintenance management (Table 4.A1).

China requires that all wind turbines meet domestic government standards and test certifications in order to be eligible for domestic wind-energy projects. Since January 2011, the National Energy Administration requires all wind turbines to have a national test certificate to obtain a construction permit or join a tendering process. Those test certification procedures create hindrance for foreign wind-turbine manufacturers (European Commission, 2012). Since the Chinese standards differ from international ones, foreign turbine manufacturers must adjust their wind-turbine models to conform to Chinese standards, which can hamper FDI using foreign technology (KTH, 2011).

Table 4.A1 **Summary of 18 wind power technical standards issued in 2011**

S/N	Standard Type	Standard No.
1	Technical specifications for grid connection design of large scale wind farms	NB/T 31003-2011
2	Guidelines for vibration condition monitoring and diagnosis of wind turbine generator systems	NB/T 31004-2011
3	Method for Testing Quality of Electric Energy of Wind Farms	NB/T 31005-2011
4	Technical Standards for Steel Structure Corrosion Resistance of Offshore Wind Farm	NB/T 31006-2011
5	Charging Standard for Investigation and Design of Wind Farm Project	NB/T 31007-2011
6	Quota of Budgetary Estimate for Offshore Wind Farm Project	NB/T 31008/2011
7	Compilation Rules and Charging Standard for Budgetary Estimate of offshore Wind Farms	NB/T 31009/2011
8	Quota of Budgetary Estimate of On-land Wind Farms	NB/T 31010/2011
9	Compilation Rules and Charging Standard for Design Budgetary Estimate of On-Land Wind Farms	NB/T 31011-2011
10	Manufacture and Technical Specifications for Permanent Magnet Type Wind Turbine Generators	NB/T 31012-2011
11	Manufacture and Technical Specification for Doubly Fed Type Wind Turbine Generators	NB/T 31013-2011
12	Manufacture and Technical Specifications for Converters of Doubly Fed Type Wind Turbine Generators	NB/T 31014-2011
13	Manufacturing and Technical Specifications for Converters of Permanent Magnet Type Wind Turbine Generators	NB/T 31015-2011
14	Technical Specifications for Battery Energy Storage Power Control System	NB/T 31016-2011
15	Technical Specifications for main Control Systems of Doubly Fed Type Wind Turbine Generator Units	NB/T 31017-2011
16	Technical Specifications for Electric Pitch Control System of Wind Turbine Generator Units	NB/T 31018-2011
17	Corona-Resistant Polyamide Film-Backed Mica Paper Tapes with Glass Fabric for Coil Insulation on Wind Turbine Generators.	NB/T 31019-2011
18	Corona-Resistant Polyamide Film for Turn-to-Turn Insulation on Wind Turbine Generators	NB/T 31020-2011

Intellectual property

Obtaining the Chinese intellectual-property certificate associated with advanced wind-turbine designs (hence using Chinese standards) is a key priority for the Chinese government in its support for Chinese wind-turbine manufacturers. In 2006, the Provisional Measures for the Accreditation of National Indigenous Innovation stated that products made with Chinese intellectual property could qualify for priority in government procurement (Global Energy Network Institute, 2010). While the Chinese government officially ended this practice in December 2011, the policy change did not apply to purchases made by China's state-owned enterprises (SOEs), who owned 81% of installed wind-energy capacity in 2012. In 2012, among the 1 300 companies who had invested in or built wind power development projects, 1 000 were controlled by SOEs (China Daily, 2013).

Grid connection requirements

Since June 2012, the technical professional standard "Technical Specifications for Grid-Connection Design of Large Scale Wind Farm" requires all turbines to be equipped with low-voltage ride-through capability (LVRT), which allows wind generators to operate through periods of lower-than-normal grid voltage. This technical standard (NB/T 31003-2011) differs from the international IEC standard for wind turbines (IEC 61400-2) (GWEC, 2014).

Environmental impact assessments

No specific technical restrictions have been found with respect to EIA for wind development that favours local investors over foreign investors.

Country factsheet of India

Summary of findings

Technical requirements in wind energy in India do not pose restrictions for foreign investors. Wind turbines installed in India are required to comply with Indian standards and certifications, which are aligned with international standards adapted to Indian geo-physical conditions. Major stakeholders have expressed an interest in establishing a certification facility in India. Testing services are market driven, so developers often choose the Center for Wind Energy Technology (CWET) over various international testing laboratories due to competitive pricing. The CWET works closely with international experts on standards. The analysis finds that environmental permitting is burdensome and subject to multiple local authorities and communities which could potentially increase transaction costs for both international and domestic investors.[8]

Market outlook

The wind power market in India began its initial growth in the mid-1990s and saw a second resurgence in the early 2000s. By 2011, India had the fifth largest installed wind capacity base globally (20 150 MW). In 2013 alone, 1 729 MW of wind power were installed. Wind energy accounts for 70% of India's renewable-power generation, excluding hydroelectricity. The total installed capacity is expected to reach almost 31.4 GW by 2015, growing to 59 GW by 2020 and 124 GW by 2030.

Several international companies with subsidiaries in India are sourcing more than 80% of their components from Indian component manufacturers. Leading global manufacturers like Enercon, RRB Energy, Suzlon, Vestas and newer entrants like Gamesa, GE, Siemens, and Regen Powertech, have set up production facilities in India (GWEC, 2012a). The shift in business model (i.e. greater flexibility) will influence independent power producers' procurement of turbines with better performance, so they do not have to rely on local original equipment manufacturers (OEMs). This will open opportunities for additional foreign OEMs to enter the market.

Administrative procedures for wind development

In the absence of regulatory restrictions on land acquisition, preferential administrative treatment for local companies can hinder the implementation of foreign investment projects. However, no specific technical restrictions have been found with respect to administrative procedures for wind development that favour local investors over foreign investors in India.

Wind turbines standards

Indian design certification

In India, the competitive tendering process and the lack of turbine standards or production requirements cause poor performance of several early projects. In 2003, however, certification of design and performance became mandatory. This reduced concerns about sub-standard technology and implementation. The key lesson learned from the Indian experience is that competitive tendering must be accompanied by robust technological standards to avoid a downward pricing pressure leading to poor quality projects.

Foreign wind-turbine manufacturers typically partner with domestic ones, either through joint ventures or technology-transfer arrangements. These manufacturers (i.e. Enercon, Gamesa, GE Wind, RBB Energy, Suzlon and Vestas) often supply wind turbines and equipment that is certified by internationally accredited certification bodies. These certificates are for instance based on European site conditions and approved technical criteria. Turbines installed in India then need to undergo minor changes to adapt to Indian conditions. All major stakeholders have expressed an interest in establishing a certification facility in India (GWEC, 2012a).

The CWET is a national institution that provides wind resource assessments, standardisation and certification support to the wind industry in India. Developers can use IEC/ISO standards as well as TAPS-2000 standards for the testing and certification of wind turbines. TAPS-2000, the Indian certification scheme for wind turbines (amended in April 2003), has been prepared by the CWET in line with international standards tailored to the specific Indian conditions. The scheme was issued and approved by the Ministry of New and Renewable Energy (MNRE). Similarly to the IEC standards, it comprises three categories including design, type test and manufacturing of system modules. The scheme is aligned with international standards taking into account Indian conditions. MNRE has established the Standards and Certification (S&C) Unit which works closely with international experts on standards.

Testing reports are accepted by international certification bodies. If the CWET does not carry out this service, another international body can carry out this service in India. Meanwhile, the preparation of Indian standards is underway. The CWET is working in close coordination with India's National Standard Body (BIS). The CWET reviewed the draft of Indian standards and supported BIS on various issues for coordination with IEC-TC 88 standards for wind turbines. India participated in IEC-TC 88 committee meeting. The CWET has been closely interacting with international research institutes such as: Riso Denmark Research Center for wind forecasting and certification; National Renewable Energy Laboratory of the United States for joint research work; and ETP Scotland for offshore wind power.

Grid connection requirements

The study did not identify restrictions for foreign investors in terms of grid connection requirements. Learning process is ongoing. However, in an effort to shore up funds for state utilities and also improve grid stability, the Central Electricity Regulatory Commission (CERC) has made it mandatory for wind-power projects greater than 10 MW to forecast their generation for the next day on a quarter-hourly basis, or to face a penalty fine paid to state utilities. This is likely to affect independent power producers (IPPs) with their larger scale wind-energy projects and could prove technically challenging. The new forecasting requirements are now being contested in court. The lack of a unified grid that integrates local, regional and national grids continues to pose a major challenge for both international and domestic investors (MAKE Consulting, 2011).

Environmental assessment procedures

Regional specification for consent to establish and to operate

Environmental permitting in India is burdensome and subject to multiple local authorities and communities, which could potentially increase transaction costs for both international and domestic investors. The State Pollution Control Boards (SPCBs) are the competent authorities to grant "consent to establish" (CTE) and "consent to operate" (CTO) for wind- and solar-energy projects, after assessing environmental impacts and the design of pollution control installation, and verifying compliance with these conditions. A CTO is issued with emission and effluent limits based on industrial sector-specific standards. Some states (e.g. Gujarat) issue consolidated consents for air and water pollution and hazardous waste, based on Common Consent Applications (CCAs). Other states such as Chhattisgarh issue separately water and air consents, as well as waste management authorisations.

Country Factsheet of South Africa

Summary of findings

In South Africa, the analysis concludes that there is no significant technical restriction specific to foreign investors in wind energy. Most investments are done by foreign investors, and international standards for wind turbines apply to South Africa. There are strict rules, for grid connection in particular, yet these rules do not seem intended to discourage foreign companies.[9]

Market outlook

South Africa aims to reach 10% of electricity generation capacity from renewable energy by 2030 (GWEC, 2012a). The Integrated Resource Plan calls for a total of 8.4 GW of wind energy by 2030. South Africa is the country with the highest wind potential in the sub-Saharan region. Since 2011, three bidding rounds have taken place in South Africa, which planned more than 2 GW of new wind-energy capacity. Mainstream Renewable Power is the largest developer on the South African market, with more than 490 MW secured. Vestas has the biggest share of the manufacturers' market, with more than 400 MW.

Wind turbine standards

Certificate of proven technology

To obtain permits for developing a wind farm, bidders must provide a valid certificate of proven technology issued by an international certification body, including Det Norske Veritas (DNV), Germanischer Lloyd, (GL), Technischer Überwachsungs-Verein (TÜV SÜD) or German Wind Energy Institute Offshore and Certification Center (DEWI-OCC). These requirements do not seem to represent a constraint neither for foreign nor for domestic manufacturers, since both are able to get their turbines certified onsite directly in South Africa. The International Electro-technical Commission (IEC) code 61400 governs wind turbine standards in South Africa (Electrical Power Research Institute, 2012).

Grid connection requirement

Grid connection standards in South Africa differ from international standards, but there is no attempt to discourage international investors. The Grid Connection Code introduced by the National Energy Regulator of South Africa (NERSA) in 2012 lists all requirements power plants have to comply with in order to be connected to the grid. These requirements differ from European ones and are usually stricter. For example, special requirements for fault-ride-through[10] may be challenging for certain models of wind turbines. Power plants must also be designed to withstand sudden voltage phase jumps of up to 40° at the point of connection without disconnecting or reducing its output, and to withstand voltage drops and peaks without disconnecting. The nominal frequency of the National Integrated Power System (NIPS) is 50 Hertz (Hz). The power plants are disconnected from the grid only if their frequency is higher than 52 Hz for more than 4 seconds, or less than 47.0 Hz for longer than 200 milliseconds.

Environmental assessment procedures

Environmental requirements are very strict with respect to geophysical requirements such as water consumption. They affect both international investors and domestic ones. Given water scarcity in the region, water related issues are perceived as very important. For instance, a written confirmation of water allocation for all the water consumption needs of the project has to be provided, which could increase transaction costs and delay the process of gathering documentation required to participate in the bidding process.

Notes

1. ICTSD (2013c); Interview with Brazilian Wind Energy Association (ABBeolica) (2014); Interview with international wind energy developer investing in Brazil (2014); IRENA and GWEC (2013).

2. CSA (2008, 2012); GWEC (2013); ICTSD (2013c, 2007); Stepp and Atkinson (2012); OECD (2015); OPA (2013); World Resources Institute (2009).

3. For a list of projects in Canada please see: http://www.canwea.ca/pdf/CanWEA-Installedcap-February-2014.pdf

4. Ontario's Long Term Energy Plan: http://www.energy.gov.on.ca/en/ltep/.

5. The North American Electricity Reliability Council.

6. China General Certification Center (2013); Cornell University ILR School (2011); Covington & Burling LLP (2014); GWEC (2013); GWEC (2012b); ICTSD (2013c); Stepp and Atkinson (2012); Ministry of Finance (2011); National Energy Administration (2013); Navigant Research (2013); OECD (2009, 2010c, 2015); and World Resources Institute (2009).

7. NB is the code referring to Energy professional standards.

8. Gomathinayagam (2013); Government of India (2008); Ministry of Law and Justice (2003).

9. Greenpeace (2011); Republic of South Africa Department of Energy (2013a, b, c, 2002); ICTSD (2013a, b); Chadbourne and Parke LLP and Yasse Yaqub (2013); Windpower Monthly (2014); GWEC (2012a).

10. Fault-ride-through is the design capability of generators to withstand faults from the grid without disconnecting, thus avoiding propagation of the failure (generators "ride through the fault").

References

AFASE (2013), *Open Letter to the EU Commissioner for Trade*, Alliance for Affordable Solar Energy, Brussels.

Aggarwal, V.K. and S.J. Evenett (22 October 2012), "The sad truth behind growing clashes at the WTO", Harvard Business Review Blog Network, http://blogs.hbr.org/2012/10/the-sad-truth-behind-the-growi/.

Ang, G. and V. Marchal (2013), "Mobilising private investment in sustainable transport infrastructure: The case of land-based passenger transport", *OECD Environment Working Paper*, No. 56, OECD Publishing, Paris, http://doi.org/10.1787/5k46hjm8jpmv-en.

Australian government (2007), *Renewable Energy Law in China – Issues Paper*, Renewable Energy Generators Australia Ltd.

Australian Productivity Commission (2010), "Bilateral and regional trade agreements", *Research Report*, No. 43, Productivity Commission, Canberra.

Bahar, H., J. Egeland and R. Steenblik (2013) "Domestic incentive measures with possible trade implications", *OECD Trade and Environment Working Papers*, No. 2013/01, OECD Publishing, Paris, http://dx.doi.org/10.1787/5k44srlksr6f-en.

Capobianco, A. and H. Christiansen (2011), "Competitive neutrality and state-owned enterprises: Challenges and policy options", *OECD Corporate Governance Working Papers*, No. 1, OECD Publishing, Paris, http://dx.doi.org/10.1787/5kg9xfgjdhg6-en.

CAWEA (2014), Market Overview of Canadian Provinces, Canadian Wind Energy Association.

CEPR (2012), *Débâcle: The 11th GTA Report on Protectionism*, Centre for Economic Policy Research, London.

CERDI (2010), "Environmental law in Brazil- analysis of environmental licensing of wind power plants in permanent preservation areas", *the International Conference: Environment and Natural Resources Management in Developing and Transition Economies*, CERDI, Clermont Ferrand.

Chadbourne & Parke LLP and Y. Yaqub (2013), "South Africa: Lessons from projects financed to date", *Project Finance NewsWire*, Chadbourne & Parke LLP.

China General Certification Center (CGC) (2013), *Introduction on Certification and Testing of Wind Turbine in China*, China General Certification Center, June 2013.

China Daily (2013), "SOEs dominate China Wind Power Projects", China Daily, Beijing.

Christopher J. Ferguson-Martin and Hill, Stephen D (2011), "Accounting for variation in wind deployment between Canadian provinces", *Energy Policy*, Vol. 39, Issue 3.

Cimino and Hufbauer (2014), "Trade remedies targeting the renewable energy sector," *Green Economy and Trade. Ad hoc Expert Group 2: Trade Remedies in Green Sectors: the Case of Renewables*, UNCTAD, Geneva, http://unctad.org/meetings/en/SessionalDocuments/ditc_ted_03042014Petersen_Institute.pdf.

Cornell University ILR School (2011), *China and the United Stated, A Comparison of Clean energy Programs and Policies*, CRS Report for Congress, Congressional Research Service, Washington, DC.

Covington & Burling LLP (2014), "Measures and practices restraining foreign investment in China", Prepared for the European Commission Directorate-General for Trade; Covington & Burling LLP, http://trade.ec.europa.eu/doclib/docs/2014/august/tradoc_152739.08.10.pdf.

CSA (2012), *CSA Guide to Canadian Wind Turbine Codes and Standards*; Canadian Standards Association.

CSA (2008), *CSA Guide to Canadian Wind Turbine Codes and Standards*; Canadian Standards Association.

De Gucht, K. (2012), "A trade defence system for today's global economy", speech of the European Commissioner for Trade, Brussels, 10 May 2012.

De Gucht, K. (2010), “Speaking points: Anti-dumping cases state of play and perspectives”, oral questions in the European Parliament, Strasbourg, 24 November 2010.

Electrical Power Research Institute (2012), Power Generation Technology Data for Integrated Resource Plan of South Africa, July 2010.

European Commission (EC) (2012), Ninth Report on Potentially Trade Restrictive Measures, 2012.

Exim Bank (2011), "Ex-Im Bank Announces $ 9.2 Million Loan to Support Abound Solar Inc. Exports", press release, Export-Import Bank of the United States, June 18, 2011.

Ghosh, A. and R. Meléndez-Ortiz (2013), “Want clean energy? Avoid trade disputes”, *Business Standard*, Business Standard Ltd.

Ghosh, A. and H. Gangania (2012) *Governing Clean Energy Subsidies: What, Why, and How Legal?,* ICTSD Global Platform on Climate Change, Trade and Sustainable Energy, International Centre for Trade and Sustainable Development, Geneva.

Gibson, S. (2014), “Trade wars over PV panels could boost prices”, *GreenBuildingAdvisor.com.*

Global Energy Network Institute (GENI) (2010), *Chinese Policies Hurt US Renewable Industries*, Dewey & LeBeouf LLP, Washington, DC.

Global Trade Alert (2013), *Global Trade Alert Statistics*.

Global Trade Alert (2012a), “EC: Initiation of antidumping investigation on crystalline silicon photovoltaic modules imported from China”, published 19 November 2012.

Global Trade Alert (2011), “China: Provisions on implementing the security review system for M&A of local enterprises by foreign investors”.

Golub, S. S., C. Kauffmann and P. Yeres (2011), “Defining and measuring green FDI: An exploratory review of existing work and evidence”, *OECD Working Papers on International Investment*, 2011/02, OECD Publishing, Paris, http://dx.doi.org/10.1787/5kg58j1cvcvk-en.

Gomathinayagam, S. (2013) “Indian wind turbine testing, standards and certification and international collaborations”.

Goodrich, A. et al. (2011), “Solar PV Manufacturing Cost Analysis: U.S. Competitiveness in a Global Industry”, National Renewable Energy Laboratory (“NREL”).

Goossens, E. (2012), “Average U.S. Wind Farm Creates 1,079 Jobs, Report Finds”, *Bloomberg News.*

Government of India (2008), *National Action Plan on Climate Change.*

Greenpeace (2011), *Powering the Future: Renewable Energy Roll Out in South Africa.*

GTM (2012), "PV Technology, Production and Cost: 2012-2016 Outlook", GTM Research.

GTM (2011), *U.S. Solar Energy Trade Assessment 2011: Trade Flows and Domestic Content for Solar Energy-Related Goods and Services in the United States*, Greentech Media.

Guérin, E. and J. Schiavo (2011), "Chinese renewable energy and technology policies: Legal compatibility with WTO rules & economic interactions with other countries' climate and industrial policies", *IDDRI*, No. 02/11, the Institute for Sustainable Development and International Relations, Paris.

GWEC (2014), "Global installed wind power capacity - regional destribution", www.gwec.net/wp-content/uploads/2012/06/01_glob-inst-wp-cap-reg-dist.jpg.

GWEC (2013), *Annual Report*, Global Wind Energy Council.

GWEC (2012a), *Annual Report*, Global Wind Energy Council.

GWEC (2012b), "China wind energy development update", Global Wind Energy Council.

ICTSD (2013a), "Policy dialogue at the occasion of the Clean Energy Ministerial (CEM)", International Centre for Trade and Sustainable Development-CEEW (Council on Energy, Environment and Water), New Delhi, 16 April 2013.

ICTSD (2013b), *Local Content Requirements and the Green Economy,* International Centre for Trade and Sustainable Development, Geneva.

ICTSD (2013c), *Addressing Local Content Requirements in a Sustainable Energy Trade Agreement,* International Centre for Trade and Sustainable Development, Geneva.

ICTSD (2012), "WTO panel hearing: Canada defends feed-in tariff as necessary government procurement", *Bridges Weekly Trade News Digest*, Vol. 16 No 12, International Centre for Trade and Sustainable Development, Geneva.

ICTSD (2011a), *Fostering Low Carbon Growth: The Case for a Sustainable Energy Trade Agreement*, International Centre for Trade and Sustainable Development, Geneva.

ICTSD (2011b), *Feed-in Tariffs for Renewable Energy and WTO Subsidy Rules,* Trade and Sustainable Energy Series, Issue paper 4, International Centre for Trade and Sustainable Development, Geneva.

ICTSD (2009), *Trade Flows, Barriers and Market Drivers in Renewable Energy Supply Goods: The Need to Level the Playing Field*, International Centre for Trade and Sustainable Development, Geneva.

ICTSD (2007), *An overview of Key markets, Tariffs and Non-tariff measures on Asian Exports of Select Environmental Goods* International Centre for Trade and Sustainable Development, Geneva.

IEA (2012a), *Tracking Clean Energy Progress: Energy Technology Perspectives 2012 excerpt as IEA input to the Clean Energy Ministerial,* OECD/IEA, Paris, www.iea.org/media/etp/Tracking_Clean_Energy_Progress.pdf.

IEA (2012c), Energy Policies Beyond IEA Countries: Ukraine 2012, OECD Publishing, Paris, http://dx.doi.org/10.1787/9789264187450-en..

IEA (2012d), *World Energy Outlook 2012*, OECD Publishing, Paris.http://dx.doi.org/10.1787/weo-2012-en.

Interview by EWEA with Brazilian Wind Energy Association, 10 March 2014.

Interview by EWEA with COBEI (Comitê Brasileiro de Eletricidade, Eletrônica, Iluminação e Telecomunicações) (2014), National Committee IEC – Brazil, interviewed on 26 March 2014, EWEA.

Interview by EWEA with international wind energy developer investing in Brazil (2014), 24 February 2014, EWEA.

International Trade Administration (2014), "Commerce Preliminarily Finds Countervailable Subsidization of Imports of Certain Crystalline Silicon Photovoltaic Products from the People's Republic of China", Fact Sheet, U.S. Department of Commerce.

IRENA (2013a), *Renewable Energy and Jobs* International Renewable Energy Agency, Abu Dhabi.

IRENA (2013b), *Renewable Energy Auctions in Developing Countries*, International Renewable Energy Agency, Abu Dhabi.

IRENA (2011), *Renewable Energy Jobs: status, prospects and policies, International Renewable Energy Agency Working Paper*, IRENA, Abu Dhabi.

IRENA and GWEC (2013), "Brazil", *30 Years of Policies for Wind Energy*, International Renewable Energy Agency and Global Wind Energy Council.

ITA (2010), "Renewable energy market assessment report: Indonesia", Manufacturing and Services Competitiveness Report, International Trade Association, Washington, DC.

Jha, V. (2013a), *Removing Trade Barriers on Selected Renewable Energy Products in the Context of Energy Sector Reforms: Modelling the Environmental and Economic Impacts in a General Equilibrium Framework*, ICSTD, Geneva.

Kalinova, B., A. Palerm and S. Thomsen (2010), "OECD's FDI Restrictiveness Index: 2010 Update", OECD Working Papers on International Investment, No. 2010/03, OECD Publishing, Paris, http://dx.doi.org/10.1787/5km91p02zj7g-en.

Kasteng, Jonas (2014), "Trade Remedies on Renewables: Intermediate and Long-Term Solutions", presentation at UNCTAD Clean energy and Trade Ad hoc Expert Group 2 on Trade Remedies in Green Sectors: The Case of Renewables, Geneva, 3-4 April 2014.

Kasteng, Jonas. (2013), "Trade remedies on clean energy: A new trend in need of multilateral initiatives", think piece for the E15 Expert Group on Clean Energy, ICTSD, Friedrich Ebert Stiftung and Chatham House.

Kommerskollegium (2013a), "Effects on trade and competition of abolishing anti-dumping measures: The European Union experience", Kommerskollegium (Swedish National Board of Trade), Stockholm, www.kommers.se/Documents/dokumentarkiv/publikationer/2013/rapporter/report-effects-on-trade_webb.pdf.

Kommerskollegium (2013b), "Targeting the environment: Exploring a new trend in the EU's trade defence investigations", Kommerskollegium (Swedish National Board of Trade), Stockholm.

Konings, J. and H. Vandenbussche (2010), "Antidumping hurts exporters: Firm-level evidence from France", European Research Workshop in International Trade (ERWIT) held jointly with the 2nd EFIGE Scientific Workshop and Policy Conference, Rome, 16-18 June 2010.

Kowalski, P., et al. (2013), "State-Owned Enterprises: Trade Effects and Policy Implications", OECD Trade Policy Papers, No. 147, OECD Publishing, Paris, http://dx.doi.org/10.1787/5k4869ckqk7l-en.

Krulewitz, A. and S. Mehta (2011), "The direction of the vertically integrated: Downstream", Greentech Media.

KTH (2011), "Wind power in China, a study of the Chinese distributed market through the example of Guangdong province", Bachelor of Science thesis, KTH Industrial Engineering and Management, Stockholm.

Latin American Herald Tribune (2014), "Brazil limits land sales to foreigners", Latin American Herald Tribune, Caracas, 2014.

Lincicome, S. (2012), "Countervailing calamity: How to stop the global subsidy race", *Policy Analysis,* No. 710, CATO Institute, Washington, DC.

MAKE Consulting (2011), "Policy shift impacts supply dynamics in India", Flash note, MAKE Consulting.

Martin, C. (2014), "Solar thin-film panels reach 11-month high on trade war", *Bloomberg News*.

Martin, C. (2012), "JinkoSolar soars with $1 billion in backing from China", *Bloomberg,* 7 December 2012.

Mercom Capital Group (2011), "Loans and credit agreements involving Chinese banks to Chinese solar companies since Jan 2010", Mercom Capital Group.

Miner, J. (2013), "Despite end of domestic content rules, CanWEA president Robert Hornung says Ontario's capacity set to grow in 2014", *The London Free Press*, 12 December 2013.

Ministry of Economy, Trade and Industry (METI) (2013), "Statement by minister Motegi, Minister of Economy, Trade and Industry on circulation of the WTO Appellate Body report on certain local content requirements under the feed-in tariff program in Ontario, Canada", 7 May 2013.

Ministry of Finance (2011), "The interim measure of renewable energy power price additional subsidy funds management", http://en.cnrec.info/policy/domestic/electricity/2012-05-09-1.html.

Ministry of Law and Justice (2003), *Electricity Act*, No. 36 of 2013, http://faolex.fao.org/docs/pdf/ind82256.pdf.

Ministry of Mines and Energy (2014), *Regulation 34/2014*.

National Energy Administration (2013), "Notice on integrating and accommodating wind power via the energy collective".

Navigant Research (2013), *BTM Wind Report, World Market update 2012*, Navigant Research.

OECD (2015, forthcoming), "Emerging policy issues: Localisation barriers to trade, for an application of the trade model on quantitative restrictions to trade", OECD Trade Policy Working Paper, OECD Publishing, Paris.

OECD (2015), *Policy Guidance for Investment in Clean Energy Infrastructure: Expanding Access to Clean Energy for Green Growth and Development,* OECD Publishing, Paris, http://dx.doi.org/10.1787/9789264212664-en.

OECD (2013a), *OECD Code of Liberalisation of Capital Movements,* OECD Publishing, Paris, http://www.oecd.org/daf/inv/investment-policy/CapitalMovements_WebEnglish.pdf.

OECD (2013b), *Inventory of Estimated Budgetary Support and Tax Expenditures for Fossil Fuels 2013*, OECD Publishing, Paris, http://dx.doi.org/10.1787/9789264187610-en.

OECD (2013c), *Interconnected Economies: Benefiting from Global Value Chains*, OECD Publishing, Paris, http://dx.doi.org/10.1787/9789264189560-en.

OECD (2013d), *OECD Glossary of Statistical Terms*, http://stats.oecd.org/glossary/.

OECD (2013e), "OECD-WTO Database on Trade in Value-Added FAQs: Background Note", OECD, http://www.oecd.org/sti/ind/TIVA_FAQ_Final.pdf.

OECD (2013f), "The Climate Challenge: Achieving Zero Emissions", lecture by the OECD Secretary-General, Mr. Angel Gurría, London, 9 October 2013.

OECD (2013g), "OECD roundtable consultation on achieving a level playing field for international investment in solar and wind energy", Paris, 6 December 2013, Summary report.

OECD (2010a), "State owned enterprises and the principle of competitive neutrality ", OECD Competition Committee, Policy Roundtables, OECD Publishing, Paris.

OECD (2010b), "Electricity: Renewables and Smart Grids", OECD Competition Committee, Policy Roundtables, OECD Publishing, Paris.

OECD (2010c), "Green growth and the crisis", brief, OECD.

OECD (2010d), "Trade and the economic recovery: Why open markets matter", brief, OECD.

OECD (2010e), "Competition, state aids and subsidies", OECD Global Forum on Competition, Policy Roundtables, OECD Publishing, Paris.

OECD (2009), *Facilitating Trade in Selected Climate Change Mitigation Technologies in the Energy Supply, Buildings and Industry Sectors* OECD Publishing, Paris.

OECD (2008a), *Proportionality of Security-Related Investment Instruments: A Survey of Practices*, OECD Publishing, Paris.

OECD (2008b), *OECD Benchmark Definition of Foreign Direct Investment,* Fourth Edition, OECD Publishing, Paris.

OECD (2006), *Policy Framework for Investment,* OECD Publishing, Paris.

OECD (2005), *OECD Guidelines on Corporate Governance of State-Owned Enterprises,* OECD Publishing, Paris.

OECD (2004), *OECD Principles of Corporate Governance,* OECD Publishing, Paris.

OECD (2003), "Foreign Direct Investment Restrictions in OECD Countries", in *OECD Economic Outlook: Special Focus on Foreign Direct Investment,* OECD Publishing, Paris.

OECD (2002), *Competition Issues in the Electricity Sector*, OECD Publishing, Paris.

OECD, WTO and UNCTAD (2014), "Report on G20 trade and investment measures (mid-November 2013 to mid-May 2014)".

OECD, WTO and UNCTAD (2013a) "Implications of global value chains for trade, investment, development and jobs", prepared for the G-20 Leaders Summit at Saint Petersburg, September 2013.

OECD, WTO and UNCTAD (2013b), "Report on G20 trade and investment measures (mid-October 2012 to mid-May 2013)".

OPA (2013), "Feed-in tariff program: FIT rules", Ontario Power Authority, available at http://fit.powerauthority.on.ca/what-feed-tariff-program

REN21 (2014), *Renewable 2014 Global Status Report,* Renewable Energy Policy network for the 21st Century.

REN21 (2013), *Renewables 2013 Global Status Report,* Renewable Energy Policy network for the 21st Century.

Rennkamp, B., and F. Fortes Westin (2013), *Feito no Brasil? Made in South Africa? Boosting technological development through local content policies in the wind energy industry,* Energy Research Centre (ERC), University of Cape Town, Cape Town.

Republic of South Africa Department of Energy (2013a), "Renewable energy independent power producer procurement program", www.ipprenewables.co.za/.

Republic of South Africa Department of Energy (2013b), *Integrated Resource Plan for Electricity.*

Republic of South Africa Department of Energy (2013c), "Request for qualification and proposals for new generation capacity under the IPP Procurement program volume 3: Technical requirement".

Republic of South Africa Department of Energy (2013d), "Request for qualification and proposals for new generation capacity under the IPP Procurement program Part B: Qualification criteria".

Republic of South Africa Department of Energy (2002), *White Paper on the Promotion of Renewable Energy and Clean Energy Development Part one – Promotion of Renewable Energy*, Department of Minerals And Energy.

SolarServer (2014a), "IHS: Tariffs will cause U.S. PV module prices to rise, but no supply interruptions", *SolarServer*, 18 April 2014,

SolarServer (2014b), "PVX spot market price index solar PV modules".

Steenblik, R., T. Matsuoka and J. Hight (2009), "Facilitating trade in selected climate-change mitigation technologies in the electricity generation and heavy industry sectors", document of Joint Working Party on Trade and Environment, OECD, Paris.

Steenblik, R. and T. Matsuoka (2008), "Facilitating trade in selected climate-change mitigation technologies in the electricity generation and heavy industry sectors", document of Joint Working Party on Trade and Environment, OECD, Paris.

Stepp, M. and R.D. Atkinson (2012), "Green mercantilism: Threat to the clean energy economy", ITIF.

Stone, S (2014), "Localisation barriers to trade, for an application of the trade model on quantitative restrictions to trade", *OECD Trade Policy Working Paper,* Emerging Policy Issues, OECD Publishing, Paris.

The Solar Foundation (2014), *National Solar Jobs Census 2013: The Annual Review of the U.S. Solar Workforce,* The Solar Foundation, Washington, DC.

The Solar Foundation (2011), *National Solar Jobs Census 2011: A Review of the U.S. Solar Workforce,* The Solar Foundation, Washington, DC.

The Times of India (2014), "Renewable energy ministry opposes anti-dumping duties on imported solar modules", The Times of India, 7 May 2014.

UNCTAD (2010), *World Investment Report 2010: Investing in a Low-Carbon Economy,* United Nations Publication, New York.

UNEP (2014), *South-South trade in renewable energy: A trade flow analysis of selected environmental goods*, 104 p.

USITC (2011), "U.S. solar photovoltaic (PV) cell and module trade overview", USITC Executive Briefings on Trade, United States International Trade Commission, Washington, DC.

Vandenbussche H. and M. Zanardi (2010), "The Chilling Trade Effects of Antidumping proliferation", *European Economic Review*, Vol. 54, 760–777.

Wesoff, E. (2013), "Rest in Peace: The list of deceased solar companies", Greentech Media.

Windpower Monthly (2014), "Plans for growth and jobs in South Africa become reality", Windpower Monthly.

World Bank (2013), *Investing Across Borders database,* http://iab.worldbank.org/.

World Resources Institute (2009), *It should be a Breeze: Harnessing the Potential if Open Trade and Investment Flows in the Wind Energy Industry*, December 2014

WTO (World Trade Organisation) (2014a), *WTO Dispute Settlement Gateway*.

WTO (2014b), "Joint statement regarding trade in environmental goods", Davos, 24 January 2014.

WTO (2014c), "Rules of origin", http://www.wto.org/english/tratop_e/roi_e/roi_e.htm.

WTO (2014d), *Agreement on Subsidies and Countervailing Measures* ("SCM Agreement").

WTO (1994), *Agreement on Technical Barriers to Trade* ("TBT Agreement").

Wu, M. and J. Salzman (2014), "The next generation of trade and environment conflicts: the rise of green industrial policy", *Northwestern University Law Review*, Vol. 108, No. 2, 410-474.

Xianqiang, Mao (2013), "Policy implications of the case of Chinese renewable energy industry", Ad hoc Expert Group Meeting on Domestic Requirements and Support Measures in Green Sectors: Economic and Environmental Effectiveness and Implications for Trade, Geneva, 13-14 June 2013.

Xuegong, S., G. Liyan and Z. Zheng (2012), "Market Entry Barriers for FDI and Private Investors: Lessons from China's Electricity Market", ERIA Discussion Paper Series ERIA-DP-2012-17.

Yong, H. (2014), "Trade Remedies in Green Sectors: the cases of Renewables", Ministry of Commerce, Beijing.

Glossary

Anti-dumping duty (AD)

An additional duty levied on imported goods to offset the injurious effects the dumped imports might cause to the domestic industry. Similar to a countervailing duty (CVD), an AD may not be imposed by WTO members unless it is determined that there are dumped imports, injury to a domestic industry, and a causal link between the dumped imports and the injury.

Capacity

The rated capacity of a heat or power generating plant refers to the potential instantaneous heat or electricity output, or the aggregate potential output of a collection of such units (such as a wind farm or set of solar panels). Installed capacity describes equipment that has been constructed, although it may or may not be operational (e.g., delivering electricity to the grid, providing useful heat, or producing biofuels).

Capital cost

The cost of funds used for financing a business. Cost of capital depends on the mode of financing used – it refers to the cost of equity if the business is financed solely through equity, or to the cost of debt if it is financed solely through debt.

Capital intensity

Measure of a firm's efficiency in deployment of its assets, computed as a ratio of the total value of assets to sales revenue generated over a given period. Capital intensity indicates how much money is invested to produce one dollar of sales revenue.

Capital subsidy

A subsidy that covers a share of the upfront capital cost of an asset (such as a solar water heater). These include, for example, consumer grants, rebates, or one-time payments by a utility, government agency, or government-owned bank.

Countervailing duty (CVD)

An additional levy imposed on imported goods to offset the injurious effects to the domestic industry of subsidies provided to producers or exporters by the government of the exporting country. WTO members may impose a countervailing duty whenever it is determined that there are subsidised imports, injury to a domestic industry, and a causal link between the subsidised imports and the injury.

Credit enhancement

Reducing the credit or default risk of a debt, thereby improving its credit-worthiness and increasing the overall credit rating.

Credit rating

Credit rating refers to an evaluation of individual's or company's ability to repay obligations or its likelihood of not defaulting. If credit rating is downgraded, it would increase the cost of capital due to the extent that the reward for such risky assets would be necessary as risk-premium.

"Crowding-in"

Occurs when public investment increases the marginal productivity of private capital or labour, or reduces the costs that investing firms incur and induces greater private investment than would have occurred otherwise.

"Crowding-out"

Occurs when a public intervention directly displaces the efforts of the private sector by undertaking projects the private sector would have otherwise done. Crowding out can also occur indirectly if governments use distortionary taxes to fund public investment.

Crystalline silicon (c-Si)

An umbrella term for the crystalline forms of silicon encompassing multi-crystalline silicon (multi-Si) and monocrystalline silicon (mono-Si), the two dominant semiconducting materials used in photovoltaic technology for the production of solar cells, that are assembled into a solar panel and part of a photovoltaic system to generate solar power from sunlight.

Distributed generation

Generation of electricity from dispersed, generally small-scale systems that are close to the point of consumption (e.g. off-grid systems, which are stand-alone systems for individual households or groups of consumers).

Due diligence

An investigation or audit of a potential investment prior to signing a contract.

Electricity generation

Defined as the total amount of electricity generated by power only or combined heat and power plants including generation required for own-use. This is also referred to as gross generation.

Energy services

Energy that is at disposal for end-users to satisfy their needs. This is also sometimes referred to as "useful energy". Due to transformation losses the amount of useful energy is lower than the corresponding final energy. Forms of energy services include transportation, machine drive, lighting or heat for space heating.

Export credit

A credit opened by an importer with a bank in an exporter's country to finance an export operation

Export subsidy

A government policy to encourage export of goods and discourage sale of goods on the domestic market through low-cost loans or tax relief for exporters, or government financed international advertising or R&D.

Feed-in tariff (FiT)

The basic form of feed-in policies. A guaranteed minimum price (tariff) per unit (normally kWh or MWh) is guaranteed over a stated fixed-term period when electricity can be sold and fed into the electricity network, normally with priority or guaranteed grid access and dispatch.

Feed-in tariff premium

A type of feed-in policy. Producers of electricity from renewable sources sell electricity at market prices, and a premium is added to the market price to compensate for higher costs and thus to mitigate financial risks of renewables production. Premiums are set as fixed premiums (a fixed amount is added to the market price for a certain period of time) or as flexible premiums (the exact amount is dependent from other criteria, e.g. market price, electricity demand, defined cap, defined floor).

General Agreement on Tariffs and Trade (GATT)

The General Agreement on Tariffs and Trade (GATT) was a multilateral agreement regulating international trade. According to its preamble, its purpose was the "substantial reduction of tariffs and other trade barriers and the elimination of preferences, on a reciprocal and mutually advantageous basis." It was negotiated during the United Nations Conference on Trade and Employment and was the outcome of the failure of negotiating governments to create the International Trade Organization (ITO). GATT was signed in 1947 and lasted until 1994, when it was replaced by the World Trade Organization in 1995.

Global value chains

The full range of firms' activities, from the conception of a product to its end use and beyond is called a value chain. It includes activities such as design, production, marketing, distribution and support to the final consumer. The activities in a value chain can be undertaken by a single company or divided among several (supplier) firms. They cover goods as well as services and can be concentrated at one location or spread out over different locations. The term "global value chains" was coined to reflect a strong trend towards the dispersion of value chain activities across the world. Many companies have broken up their value chains and distributed production stages across many countries; at the same time, they have outsourced parts of their value chains to external partners.

Green Growth

Fostering economic growth and development, while ensuring that natural assets continue to provide the resources and environmental services on which our well-being relies. To do this, it must catalyse investment and innovation which will underpin sustained growth and give rise to new economic opportunities.

Green industrial policies

Industrial policies with an environmental goal -or more precisely, as sector-targeted policies that affect the economic production structure with the aim of generating environmental benefits.

Grid-parity

Grid parity refers to "an energy source can generate electricity at a levelised cost that is less than or equal to the price of purchasing power from the electricity grid".

Independent power producers (IPPs)

An entity, which is not a public utility, but which owns facilities to generate electric power for sale to utilities and end users.

Intermediate goods

Goods used as inputs in the production of other goods including final goods.

Local-content requirements

Policy measures that typically require a certain percentage of intermediate goods used in the production processes to be sourced from domestic manufacturers. Local-content requirements in renewable energy policy serve as either a precondition to receive government support or an eligibility requirement for government procurement in renewable energy projects.

Merger and Acquisition (M&A)

Aspects of strategic management, corporate finance and management dealing with the buying, selling, dividing and combining of different companies and similar entities that can help an enterprise grow rapidly in its sector or location of origin, or a new field or new location, without creating a subsidiary, other child entity or using a joint venture.

Most Favoured Nation (MFN)

A status or level of treatment accorded by one state to another in international trade. The term means the country which is the recipient of this treatment must, nominally, receive equal trade advantages (e.g. low tariffs or high import quotas) as the "most favoured nation" by the country granting such treatment. In effect, a country that has been accorded MFN status may not be treated less advantageously than any other country with MFN status by the promising country.

Original Equipment Manufacturers (OEM)

A term used when one company makes a part or subsystem that is used in another company's end product. The term sometimes refers to a part or subassembly maker, sometimes to a final assembly maker, and sometimes to a mental category comprising those two in contrast to all other third party makers of parts or subassemblies from the aftermarket.

Private equity fund

A fund which use their own capital or capital raised from investors (or both) to take companies private with the aim of running them better and later taking them public or selling them at a profit.

Production tax credit

A taxation measure that provides the investor or owner of a qualifying property or facility with an annual tax credit based on the amount of renewable energy (electricity, heat, or biofuel) generated by that facility.

Public finance institutions (PFIs)

Publicly created or mandated financial institutions that have often been created to correct for the lack of market-based finance through the provision of missing financial services.

Research and development (R&D)

Activities related to the enterprise of corporate or governmental innovation. The activities that are classified as R&D differ from company to company, but there are two primary models, with an R&D department being either staffed by engineers and tasked with directly developing new products, or staffed with industrial scientists and tasked with applied research in scientific or technological fields which may facilitate future product development. In either case, R&D differs from the vast majority of corporate activities in that it is not often intended to yield immediate profit, and generally carries greater risk and an uncertain return on investment.

Regional trade agreements (RTAs)

Reciprocal trade agreements between two or more partners. They include free trade agreements and customs unions. Regional trade agreements (RTAs) cover more than half of international trade and operate alongside global multilateral agreements under the World Trade Organization (WTO).

Renewable-energy target

An official commitment, plan, or goal set by a government (at the local, state, national, or regional level) to achieve a certain amount of renewable energy by a future date. Some targets are legislated while others are set by regulatory agencies or ministries.

Renewable portfolio standards (RPS)

An obligation placed by a government on a utility company, group of companies, or consumers to provide or use a predetermined minimum renewable share of installed capacity, or of electricity or heat generated or sold. A penalty may or may not exist for noncompliance. These policies are also known as "renewable energy quotas," "renewable electricity standards," "renewable obligations," and "mandated market shares," depending on the jurisdiction.

Smart grid

Electrical grid that uses information and communications technology to co-ordinate the needs and capabilities of the generators, grid operators, end-users, and electricity market stakeholders in a system, with the aim of operating all parts as efficiently as possible, minimising costs and environmental impacts, and maximising system reliability, resilience, and stability.

State-owned enterprise (SOE)

A legal entity that is created by the government in order to partake in commercial activities on the government's behalf. A state-owned enterprise (SOE) can be either wholly or partially owned by a government and is typically earmarked to participate in commercial activities.

Solar Photovoltaics (PV)

A technology used for converting solar radiation (light) into electricity. PV cells are constructed from semi-conducting materials that use sunlight to separate electrons from atoms to create an electric current. Modules are formed by interconnecting individual solar PV cells. Monocrystalline modules are more efficient but relatively more expensive than polycrystalline silicon modules. Thin-film solar PV materials can be applied as flexible films laid over existing surfaces or integrated with building components such as roof tiles. Building-integrated PV (BIPV) generates electricity and replaces conventional materials in parts of a building envelope, such as the roof or façade. Bifacial PV modules are double-sided panels that generate electricity with sunlight received on both sides (direct and reflected) and are used primarily in the BIPV sector.

Technical barriers to trade

Technical regulations, standards, testing and certification procedures that create unnecessary obstacles to trade.

Trade remedies

Trade policy tools used to address distortions and unfair trade practices. Such remedies are divided broadly into anti-dumping duties, countervailing duties , and safeguard measures.

Trade in Value Added database

The goods and services we buy are composed of inputs from various countries around the world. However, the flows of goods and services within these global production chains are not always reflected in conventional measures of international trade. The joint OECD – WTO Trade in Value Added (TiVA) initiative addresses this issue by considering the value added by each country in the production of goods and services that are consumed worldwide.

Venture capital

An investment in a start-up business that is perceived to have excellent growth prospects but does not have access to capital markets. It is also a type of financing sought by early stage companies seeking to grow rapidly.

Note

Explanations of the terms are very condensed and may not be complete. They are not considered to necessarily reflect the official position of the OECD. Sources used include, OECD database and Investopedia.com.

ORGANISATION FOR ECONOMIC CO-OPERATION AND DEVELOPMENT

The OECD is a unique forum where governments work together to address the economic, social and environmental challenges of globalisation. The OECD is also at the forefront of efforts to understand and to help governments respond to new developments and concerns, such as corporate governance, the information economy and the challenges of an ageing population. The Organisation provides a setting where governments can compare policy experiences, seek answers to common problems, identify good practice and work to co-ordinate domestic and international policies.

The OECD member countries are: Australia, Austria, Belgium, Canada, Chile, the Czech Republic, Denmark, Estonia, Finland, France, Germany, Greece, Hungary, Iceland, Ireland, Israel, Italy, Japan, Korea, Luxembourg, Mexico, the Netherlands, New Zealand, Norway, Poland, Portugal, the Slovak Republic, Slovenia, Spain, Sweden, Switzerland, Turkey, the United Kingdom and the United States. The European Union takes part in the work of the OECD.

OECD Publishing disseminates widely the results of the Organisation's statistics gathering and research on economic, social and environmental issues, as well as the conventions, guidelines and standards agreed by its members.

OECD PUBLISHING, 2, rue André-Pascal, 75775 PARIS CEDEX 16
(97 2015 02 1) ISBN 978-92-64-22704-0 – 2015

www.ingramcontent.com/pod-product-compliance
Lightning Source LLC
Chambersburg PA
CBHW061217240926
51846CB00006B/1263
* 9 7 8 9 2 6 4 2 2 7 0 4 0 *